Aims and Scope

The series “Springer Theses” brings together a selection of the very best Ph.D. theses from around the world and across the physical sciences. Nominated and endorsed by two recognized specialists, each published volume has been selected for its scientific excellence and the high impact of its contents for the pertinent field of research. For greater accessibility to non-specialists, the published versions include an extended introduction, as well as a foreword by the student’s supervisor explaining the special relevance of the work for the field. As a whole, the series will provide a valuable resource both for newcomers to the research fields described, and for other scientists seeking detailed background information on special questions. Finally, it provides an accredited documentation of the valuable contributions made by today’s younger generation of scientists.

Theses may be nominated for publication in this series by heads of department at internationally leading universities or institutes and should fulfill all of the following criteria

- They must be written in good English.
- The topic should fall within the confines of Chemistry, Physics, Earth Sciences, Engineering and related interdisciplinary fields such as Materials, Nanoscience, Chemical Engineering, Complex Systems and Biophysics.
- The work reported in the thesis must represent a significant scientific advance.
- If the thesis includes previously published material, permission to reproduce this must be gained from the respective copyright holder (a maximum 30% of the thesis should be a verbatim reproduction from the author’s previous publications).
- They must have been examined and passed during the 12 months prior to nomination.
- Each thesis should include a foreword by the supervisor outlining the significance of its content.
- The theses should have a clearly defined structure including an introduction accessible to new PhD students and scientists not expert in the relevant field.

Indexed by zbMATH.

More information about this series at https://link.springer.com/bookseries/8790

Yunjiao Zhang

Tuning Autophagy-Inducing Activity and Toxicity for Lanthanide Nanocrystals

Doctoral Thesis accepted by
University of Science and Technology of China,
Hefei, China

Springer

Author
Dr. Yunjiao Zhang
South China University of Technology
Guangzhou, Guangdong, China

Supervisor
Prof. Wen Longping
School of Life Sciences
University of Science and Technology
of China
Hefei, China

ISSN 2190-5053 ISSN 2190-5061 (electronic)
Springer Theses
ISBN 978-981-16-8165-3 ISBN 978-981-16-8166-0 (eBook)
https://doi.org/10.1007/978-981-16-8166-0

This Springer imprint is published by the registered company Springer Nature Singapore Pte Ltd.
The registered company address is: 152 Beach Road, #21-01/04 Gateway East, Singapore 189721, Singapore

Supervisor's Foreword

Autophagy induction has been observed upon exposure of cells to a variety of nanoparticles and represents both a safety concern and an application niche for engineered nanomaterials and nanodevices. Ability to control the autophagy-inducing activity for these materials and devices is highly desirable and may conceivably be achieved through modulating material–cell interaction. In this thesis, she demonstrated that surface engineering with non-covalent high-affinity binding peptides provides a simple yet highly effective solution for the above problem. She also shows that:

- A short synthetic peptide RE-1, identified via a modified phage display approach, binds to lanthanide (LN) oxide and LN upconversion nanomaterials (UCN) with high affinity in a sequence-specific manner;
- RE-1 forms a stable coating layer on the surface of UCN nanocrystals;
- RE-1 coating reduces UCN–cell interaction resulted from both sedimentation and diffusion of the nanocrystals;
- RE-1 coating abrogates the autophagy-inducing activity and toxicity (both autophagy-dependent and autophagy-independent) for UCN nanocrystals;
- RE-1 peptide variants with specific amino acid changes exhibit differentially reduced binding capability, and correspondingly, varied ability to reduce cell interaction and autophagic response for UCN;
- Non-covalent coating with PEG did not affect cell interaction, autophagic response and toxicity for UCN;
- Addition of an arginine-glycine-aspartic acid (RGD) motif onto RE-1 creates a bifunctional peptide RE-1-RGD that enhances cell interaction, autophagic response and toxicity for UCN while retaining the nanocrystal surface coating property;
- Pre-treatment of cells with the RE-1-RGD peptide abolished the cell-interacting, autophagy-inducing and toxicity-promoting effects elicited by RE-1-RGD-coated UCN, demonstrating the RGD–integrin interaction as the underlying mechanism.

While other surface-modifying polymers such as PEG are widely employed to alter cell interaction and biological response for nanomaterials, peptides like RE-1 offer several advantages: simple non-covalent coating procedure, achieved through high-affinity peptide–nanocrystal interaction and leading to high consistency and easy scale-up; ability to "tune" the cell interaction to a desired level with the use of a series of peptide variants; and material selectivity, with the possibility of coating a particular material or device in a mixed environment. RE-1 and its peptide variants provide a versatile tool to enable the in vivo applications of lanthanide-based nanomaterials. For materials that do not need to interact with cells, the RE-1 peptide would be perfect to improve signal-to-noise ratio and reduce toxicity derived from non-specific interaction of nanoparticles with cells. On the other hand, if the cell interaction is required as in most of the applications, bifunctional or multi-functional RE-1 variant peptides can be employed by linking the RE-1 sequence with one or more specific organ-homing or cell-targeting motifs. The bifunctional peptide RE-1-RGD provided an excellent proof of concept.

Hefei, China
November 2021

Prof. Wen Longping

Abstract

Induction of autophagy, a critical cellular degradation process, has been observed upon exposure of cells to a variety of nanoparticles and represents both a safety concern and an application niche for engineered nanomaterials. Ability to control the autophagy-inducing activity for these materials and devices is highly desirable and may conceivably be achieved through modulating material–cell interaction. Here, we present a simple yet highly effective solution by surface engineering with non-covalent binding peptides. A short synthetic peptide RE-1, identified via a modified phage display approach, binds to lanthanide (LN) oxide and upconversion nanocrystals with high affinity in a sequence-specific manner, forms a stable coating layer on the surface of the nanoparticles and effectively abrogates their autophagy-inducing activity. Furthermore, RE-1 effectively abrogates the autophagy-inducing activity and toxicity for these nanocrystals through reducing sedimentation and cell interaction of nanoparticles, while RE-1 peptide variants with specific amino acid changes exhibit differentially reduced binding capability, and correspondingly, varied ability to reduce cell interaction and autophagic response. On the other hand, the addition of an arginine-glycine-aspartic acid (RGD) motif onto RE-1 creates a bifunctional coating peptide that enhances cell interaction and autophagic response for LN upconversion nanocrystals through interacting with integrins. RE-1 and its variants provide a versatile tool for tuning cell interaction to achieve the desired level of the autophagic response and may prove useful for the various diagnostic and therapeutic applications of LN-based nanomaterials and nanodevices.

Keywords Autophagy · Lanthanide (LN) nanocrystals · Phage display · Peptides · Upconversion · Toxicity · Cell interaction · Diagnostic · Therapeutic

Contents

Chapter 1
Introduction

1.1 Overview of Nanomaterials

In the past few decades, nanomaterials have caused a continuous and extensive research issue in different fields. This is because nanomaterials exhibit unique physical and chemical properties due to the nanoscale, and thus they have shown great application potentials in the fields of optics, electronics, electromagnetics, catalysis and biomedicine [1].

1.1.1 What is Nanomaterial

Nanomaterials generally refer to monocrystalline or polycrystalline materials with at least one dimension in the nanometer range (1–100 nm) among the three-dimensional spatial scales, mainly including atomic clusters, nanoparticles, nanowires, nanofilms, nanotubes and nano-solid materials, etc. [2].

1.1.2 Properties of Nanomaterials

The nanometer scale is at the intersection between the clusters of atoms and macroscopic objects. Therefore, nanomaterials have unique physical and chemical properties which cannot be found in many common macromaterials, including small size effects, surface effects, quantum size effects, macro quantum tunnel effects, and dielectric confinement effects. These basic characteristics make them useful in biomedical fields such as disease diagnosis and treatment, tissue engineering, and drug delivery, and showed great application prospects. The properties and applications of nanomaterials depend on the nanostructure, size, morphology, chemical composition, crystallinity, and surface structure. Recent years researches and studies

Y. Zhang, *Tuning Autophagy-Inducing Activity and Toxicity for Lanthanide Nanocrystals*, Springer Theses, https://doi.org/10.1007/978-981-16-8166-0_1

have focused on controlling the physical and chemical properties of nanomaterials to regulate the function of nanoparticles [3, 4].

1.1.3 Rare Earth Nanomaterials

The term "rare earth" comes from the end of the eighteenth century, when insoluble solid oxides were often called "earth". Rare earths are often isolated in the form of oxides. Minerals used to extract rare earth nanomaterials are relatively rare, and the obtained oxides are difficult to cultivate, to dissolve in water, and to separate. That is why it be named. Rare earth nanomaterials are part of the earth's crust and they are widely found in nature.

The development and application of rare earth nanomaterials has become a current research hotspot. The reason is that the materials combine the characteristics of rare earth and nano, inevitably developing comprehensive and excellent characteristics apart from those of non-rare earth nano materials and rare earth non-nano materials, with great and broad application prospects. The special properties of rare earth nanomaterials attributes to the applications in the fields including light effect, catalysis, etc., and make them widely used in contemporary communication technology, photosensitive materials, petrochemicals, computers, aerospace development, optoelectronics, metallurgy, machinery, energy, light industry, environment protection, agriculture and catalyst materials [5].

1.1.3.1 Rare Earth Metal Oxide Nanomaterials

Rare earth elements are a general term for lanthanide rare earth element groups in the periodic table of chemical elements. Rare earth elements belong to the family of IIIB in the periodic table, and include 15 elements from 57 to 71 of atomic number and 2 closely related to lanthanide elements: La, Ce, Pr, Nd, Pm, Sm, Eu, Gd, Tb, Dy, Ho, Er, Tm, Yb, Lu and Sc, Y. Among them, the rare earths with smaller atomic numbers such as Gd, Eu, are called "light rare earth (LREE)", while those with bigger atomic numbers are called "heavy rare earths (HREE)".

Rare earth metal oxides are an important part of rare earth nanomaterials. Due to its special properties, it has great application value in the field of biomedicine. Rare earth metal oxide nanomaterials used for radiation therapy to treat cancer have been widely used in China and Canada. Rare earth metal oxide nanomaterial-based fluorescent probes and nanodevices used for in vivo and in vitro diagnosis, imaging and disease treatment would have enormous potentials. However, there is not much researches on the safety and biological effects of rare earth metal oxides, with less studies on the interaction with key biological processes in cells [6].

1.1.3.2 Rare Earth Upconversion Luminescent Nanomaterials

Upconversion luminescence (UCL) refers to the phenomenon of rare earth ions absorbing two or more low-energy photons and emitting a high-energy photon. Under long-wavelength light excitation, UCL nanomaterials continuously emit light with a shorter wavelength than the excitation wavelength, usually from near-infrared light into visible light [7]. As early as the year of 1959, there was report on up-conversion luminescence. In an article published in the Physical Review Letter, Bloembergen found that infrared light of 960 nm can be used to excite polycrystals ZnS, and green luminescence of 525 nm can be observed [8]. At 1966, it was unexpectedly found by Auzel that when the Yb ions were doped in the matrix material, the visible luminescence increased by almost two orders of magnitude after Er3+, Ho3+ and Tm3+ were excited by infrared light. The idea of "upconversion luminescence" was formally proposed [9].

Upconversion luminescence is essentially a kind of anti-stocks luminescence, that is, the energy to emit is greater than the energy absorbed. Up to now, up-conversion materials are mainly solid compounds doped with rare earth elements. By utilizing the metastable energy level characteristics of rare earth elements, they can absorb multiple low-energy long-wave radiations, so that people are unable to see the conversion of infrared light to visible light. Compared with down-conversion luminescent tags such as organic fluorescent dyes and quantum dots, up-conversion luminescent nanomaterials have propoties of good light stability, strong chemical stability, narrow absorption and emission bands, long luminescence lifetime, and are not prone to photobleaching, as well as low biological toxicity, the up-conversion luminescence markers have significant advantages such as deeper light penetration depth, no interference with biological background light, and almost no damage to biological tissues due to the use of near-infrared continuous lasers as excitation sources. These characteristics of up-conversion luminescent nanomaterials overcome the shortcomings of traditional light-changing labeling materials, which are exactly the ideal markers for biological imaging [10–12]. These characteristics make UCN have unlimited application prospects in the field of biological imaging.

1.2 Prospects of Nanomaterial Biomedical Applications

1.2.1 Applications of Nanomaterials in the Fields of Biology and Medicine

Nanomaterials have the same size as biological macromolecules, and have excellent optical, electrical, and magnetic properties. They provide powerful tools for the study of cancer detection and treatment, making them useful in drug transport, disease diagnosis and treatment, and tissue engineering [3, 4]. The field of biomedicine has great application prospects and has become a hot spot in the research of new materials.

Nanomaterials have many outstanding advantages in biological and medical research and have been widely used. Firstly, nanomaterials have the characteristics of being able to pass through the interstitial space and be absorbed by cells. They also can pass through the smallest capillaries of the human body, and even pass through the blood–brain barrier. In addition, nanomaterials also have the advantages of matching the size of biomolecules and being easily transported in the human environment, which allows them to interact with cells, tissues and even organisms (especially proteins involved in key biological processes, and thus result in many special biological effects. What's more, nanomaterials have the characteristics of morphological controllability and surface modifiability, which enable the biological effects of nanomaterials to be effectively controlled artificially [13]. Nanomaterials' physical and chemical properties can be significantly improved through size control, assembly, or surface chemical modification, thereby achieving the expected biological effects. Nanomaterials have unparalleled advantages over ordinary macro-materials or small molecules due to their special nano-characteristics, and have the properties of high selectivity, high sensitivity, controllability and versatility in biomedical applications [14, 15]. Carbon nanotubes can triggered reactive oxygen reaction in vivo, and thus result in oxidative stress reaction, lipid peroxidation, mitochondrial damage, and changes in cell morphology [16, 17]. Fullerenes C_{60} can produce singlets oxygen, and has the functions of killing bacteria, protecting nerve cells, shearing DNA, and inhibiting the formation of amyloid proteins [18–20]. Polymer nanomicelles PEG-PE can effectively penetrate the inner wall of blood vessels as drug carriers, and form aggregates in cells, increasing the local drug concentration [21]. Nano TiO_2 has a significant antibacterial and bactericidal effect under ultraviolet light excitation [22]. The modified nano-gold can combine with fibrinogen and trigger the release of inflammatory cytokines under light excitation [23]. The iron oxide nanoparticles were used as a contrast agent for magnetic resonance imaging in medical magnetic resonance imaging and can also be used in diagnostic and therapeutic of tumors [24]. Eu $(OH)_3$ nanorods can effectively promote angiogenesis. Quantum dots can be used as fluorescent molecules for biological imaging and are widely used in medical diagnostics and biosensors [25]. However, since the chemical composition of quantum dots is mainly heavy metals, the clinical applications of quantum dots have been severely restricted in recent years [26]. As a fluorescent probe, rare earth-based up-conversion luminescent nanomaterials show unique charms in biomarkers and medical imaging [27, 29].

1.2.2 Application of Rare Earth Up-Conversion Luminescent Nanomaterials in the Fields of Biology and Medicine

Rare earth up-conversion luminescent nanomaterials can be converted to emit short-wavelength near-infrared light or visible light or ultraviolet light under the excitation of near-infrared light. Such materials have great R&D value and application prospects in many research fields (such as biological detection, disease diagnosis, biosensing, light display and solar cells, etc.). Uniformly dispersed, micro-sized and biocompatible up-conversion luminescent nanocrystals are applied to the imaging and continuous monitoring of cell identification and animal deep tissue. Compared with traditional down-converting luminescent materials, this material has the advantages of deep incidence, weak autofluorescence, low background noise, high sensitivity, and low toxicity [28, 29], and it is suitable for applications in the field of biomedicine with less damage to biological tissues [30]. Compared with non-linear materials such as two-photon, although the excitation process of up-conversion luminescent material is also based on two-photon or multi-photon absorption, its luminous efficiency is relatively high, and it is a new type of fluorescent material with superior performance. In addition, the material is doped with rare earth elements. By adjusting the type of the doped rare earth element and the matrix material, it can realize multicolor up-conversion visible light emission under the same infrared excitation light, and it can be used for simultaneous multi-target marking. The new materials not only have great scientific research value, but also possess broad commercial prospects and attract the favor of many commercial organizations [31]. The fluorescent materials have a wide range of applications in the fields of biochemical sensing and analysis, biological detection, drug tracking, optical data storage, optical identification, light display, environmental monitoring, solar cells and lasers, etc., and has huge market demand. The market demand for optical nanomaterials in 2010 is forecasted at USD 100 million, and it will grow at a speed of 50% in the coming years. Due to the unique properties of the up-conversion luminescent materials, they will become a replacement for traditional luminescent materials in many applications, and have an inestimable application prospects.

As the rare earth up-conversion nanomaterial is a rare earth doped fluoride nanomaterial, its low phonon energy can reduce non-fluorine emission transitions and improve luminous intensity [32, 33], which makes it standing out in many substrates such as oxides, sulfides, phosphides. It is also widely used in the fields of analysis and detection, and disease treatment [34–43]. Its advantages are particularly prominent in photodynamic therapy. Rare earth up-conversion luminescent nanoparticles are used as energy donors for photosensitizers, and doped with cyanine while covering the dioxide shoes to achieve singlet oxygen generation and energy transfer with cyanine and UCN [42]. In a living experiment, a polyethyleneimine-coated rare earth coated with folic acid and non-covalently adsorbing zinc phthalocyanine was used to upconvert luminescent nanoparticles of 50 nm. When the excitation of 980 nm is used to up-convert luminescent nanoparticles, the photosensitizer zinc phthalocyanine will generate singlet oxygen due to energy transfer, which can kill tumor cells [43]. After the mesoporous dioxide shoe-coated up-converting luminescent nanoparticles are covalently linked to the tumor-directing agent, its dual photosensitivity can effectively inhibit the growth of melanoma in tumor-bearing mice [44]. A large number of research data showed that rare earth up-conversion luminescent nanomaterials had basically replaced quantum dots and were widely used in biological imaging. Due to its unique up-conversion luminous properties, the excitation light is an infrared laser, so it has a deep light penetration depth, no biological background, fluorescent interference, and almost no damage to biological tissues. These characteristics make it particularly advantageous in deep tissue imaging and in vivo imaging. Studies have shown that the rare-earth up-conversion nanomaterials wrapped with PEI can still see clear fluorescence in tissues with a depth of 10 mm, while the quantum dots' light from the translucent skin of the rat's feet is still very weak [41].

In addition, rare earth up-conversion luminescent nanomaterials are also widely used in drug transport and gene transfection. Using light-sensitive chemical groups to protect biological molecules (DNA, RNA, proteins, drugs, etc.), under the action of different incident light, rare earth up-conversion nanomaterials can realize the remote control and release function of biological molecules. The study found that rare earth up-conversion luminescent nanomaterials can transport siRNA in cells using the fluorescence resonance energy transfer technology [45].

However, the biological applications of rare earth up-conversion luminescent nanomaterials are still in the preliminary stage of research, and there are still many problems before they are used for in vivo applications and clinical trials. The biosafety of rare earth up-conversion luminescent nanomaterials needs to be further improved. How to regulate nanomaterials to achieve safe and efficient use in biomedicine is a question worthy of further discussion.

1.2.3 Biosafety Assessment of Nanomaterials

The considerable economic benefits and technological progress brought by nanotechnology and nanomaterials have made it one of the most invested and fastest-growing areas of scientific research and technology development in developed countries. However, the unique physical and chemical properties of nanomaterials are clearly different from those of macro materials, and they cannot be detected by conventional methods and methods, which may cause pollution to human bodies and the ecological environment, thereby endangering human health. As a result, scientists have gradually recognized and valued the potential hazards of nanomaterials to human health, and carried out related research in response to concerns about possible biological safety issues. Studies on the effects and safety of nanomaterials have shown that nanomaterials are not completely beneficial, and they affect organisms at the cellular, subcellular, and even protein level. They can affect certain basic life processes such as cell division, proliferation, and apoptosis and regulate related signaling pathways, thereby generating certain biological effects at the cellular level. Nanomaterials may show certain cytotoxicity [46], and induce cell death [47]. The toxicity of nanomaterials mainly comes from the changes of physical and chemical properties that occur during the process of material conversion to nanoscale. For example, nanomaterials with enhanced stability are difficult to be degraded after being used on the surface of the human body or absorbed by the human body, and long-term accumulation in the body will affect health. At the biological level, related researches on the effects of nanomaterials and toxicity showed that nanomaterials could enter the body through the respiratory system, digestive tract, and skin to avoid the immune system's clearance and precipitate in the respiratory system to cause inflammatory reactions, leading to reduced lung clearance, chronic lung inflammation and oxidative damage. Also, nanomaterials can diffuse from the sedimentation site to surrounding tissues, penetrate the blood-gas barrier and enter the circulatory system. In addition, they can also penetrate the blood–brain barrier and the arousal nerve pathways to cause brain damage [48]. Recent studies have shown that the biosafety of nanomaterials is closely related to its induction of cell salivation.

1.3 Autophagy is a Key Biological Process for Cells to Maintain Homeostasis

1.3.1 What is Cell Autophagy

Autophagy is a life phenomenon peculiar to eukaryotic cells that uses lysosomes to degrade their damaged organelles, macromolecular substances, and long-lived proteins. It is a key cellular biological process regulated by multiple molecules. Cellular autophagy is a response of cells to stress. Through autophagy, cells clear their damaged organelles, degrade abnormal therapeutic proteins, and remove invading pathogens [49–51].

1.3.2 History of Cell Autophagy

Belgian scientist Christian de Duve was the originator of autophagy research and won the Nobel Prize in physiology and medicine because of discovery of ultrastructural of autophagosomes through electron microscope in 1950s. In 1962, Ashford and Porten came up with the phenomenon of "self-eating" in cells. And then Christian de Duve proposed the term "autophagy" firstly at the annual International Lysosome conference in 1963. But it was not until the establishment of the yeast model and the development of gene technology that people gradually learned more about autophagy. From 1960 to 1990s, the study of autophagy was mainly limited to the description of its phenomenon and morphology. In 1992, Yoshinori Ohsumi found that autophagy in yeast was similar to mammalian cells in morphology. Although autophagy was first found in mammalian cells, major breakthroughs in autophagic research were later realized in the yeast system, so this discovery laid the foundation for the genetic and molecular biology research of autophagy, and heralded the coming of the molecular times of autophagy [52]. Subsequently, Yoshinori Ohsumi found the first autophagy-related genes (ATG1) in 1998 [53]. At present, 35 autophagy-related genes have been reported [54]. In 1999, Levine of America found that autophagy genes Beclin 1 inhibited tumorigenesis, firstly revealing the correlation between autophagy and major diseases [55]. With the launch of the annual magazine Autophagy in January of 2004, researches on autophagy have entered into a stage of rapid development. The year-end outlook of the Science in 2004 listed autophagy as the top research area of the year, fully affirming the importance of autophagy research [56]. Five international conferences have been held from 2007 to 2011, reflecting the extensive development of autophagic research. As early as in 2000, scientists have predicted that autophagy would become the next "apoptosis" [57]. The figure shows the number of autophagy-related papers published each year from 1970 to 2007 searched in *Pubmed/MEDLINE*. At the same time, the number of papers recorded by "autophagy" at *Web of Science* has explosively increased in recent years, and the model was almost exactly the same as that of apoptosis-related researches, although the time has been delayed by about ten years (Fig. 1.1). In fact, by the time of March in 2012, the collected literatures on autophagy has reached 10,881 and has been cited up to more than 200,000 times (Fig. 1.2). It can be seen that autophagy, as the key biological

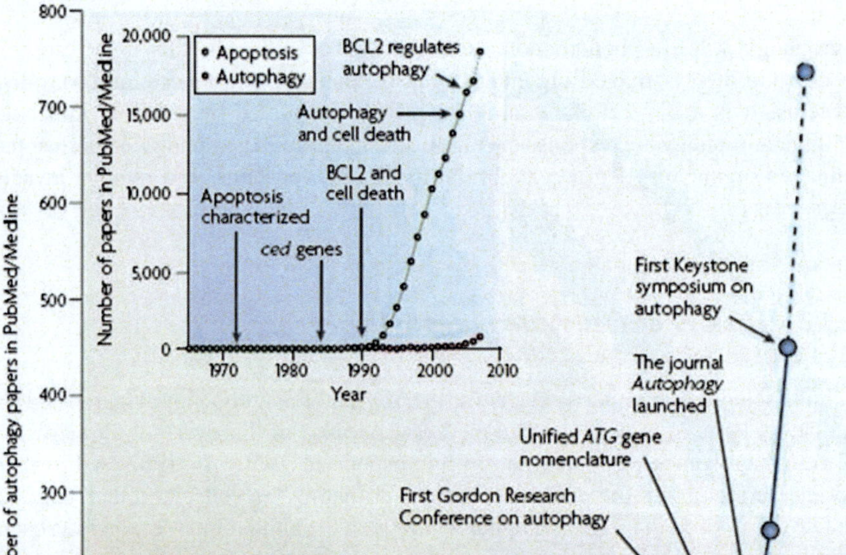

Fig. 1.1 The timeline of autophagy development

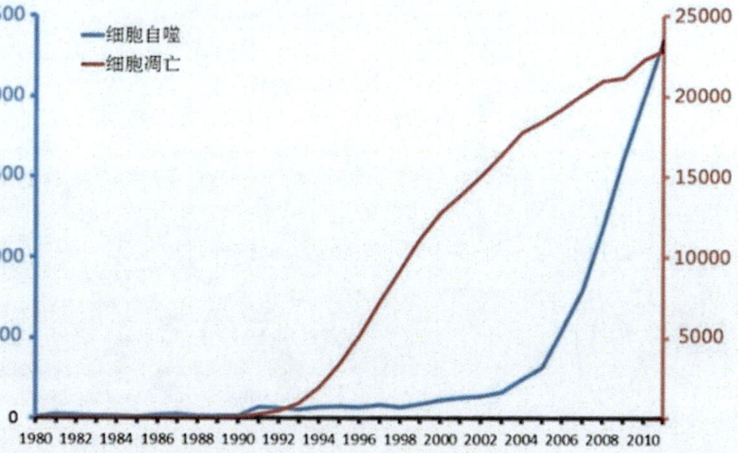

Fig. 1.2 Curve of the number of articles

process for maintaining homeostasis of cells, has become another rapidly developing field of biological research after apoptosis.

Number of papers on autophagy published annually between 1970 and 2007, retrieved by PubMed/Medline [57] (Fig. 1.2).

Number of papers on apoptosis and autophagy published annually between 1980 and 2011, retrieved by Web of Science.

1.3.3 Classification of Autophagy

Autophagy is a tightly regulated process of cell degradation of inclusions and recycling. According to the transport pathway of intracellular autophagic substrates to the lysosomes, the autophagy can be divided into three types (macroautophagy, microautophagy, and chaperone-mediated autophagy, CMA) (shown in Fig. 1.3). What we usually refer to is macroautophagy, which occurs when cells are under stress such as nutritional deficiencies and viral infections. In the process, the inclusions of the cell (including organelles, foreign substances, certain soluble proteins and protein precipitates, etc.) are enveloped by a double-layered membrane structure of non-lysosome origin, and formed to autophagosomes. Later, autophagosomes fuse with lysosomes to further become autolysosomes to degrade substrates through the action of various enzymes in lysosomes. In microautophagy, the lysosomal membrane itself is deformed by invading deformation, and directly engulfs the substrates in the cytoplasm. In the process of chaperone-mediated autophagy, the substrates are selectively transported into the lysosomes by sequence and are transported by the chaperone protein and finally are degraded. The proteins at cytoplasm are bound to the chaperone and transported to the lysosomes, which are digested by the lysosomal

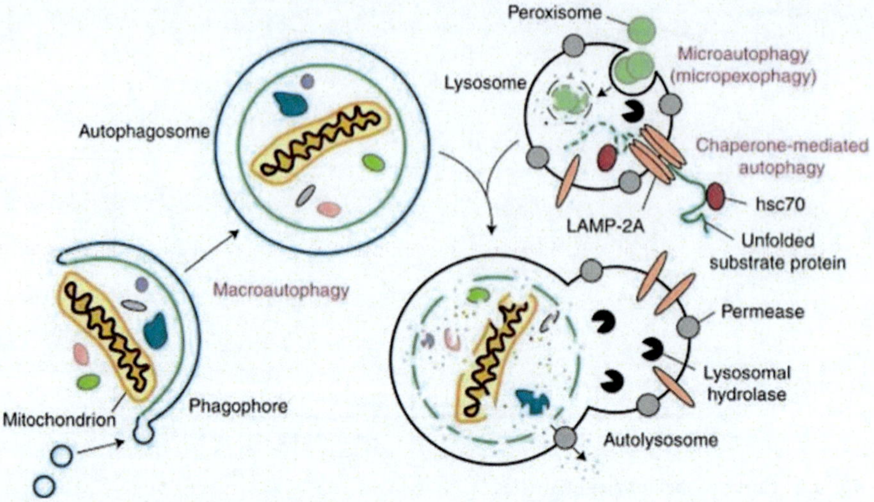

Fig. 1.3 Classification of autophagy [59]

enzyme [58]. Autophagy is a widespread degradation and recycling system found in eukaryotic cells. For example, mitophagy (mitochondrial autophagy) is the type of autophagy that clears nonfunctional mitochondria. It is the main way for cells to clear mitochondria after irreversible damage due to various factors and to prevent their contents from spreading to the entire cytoplasm. This is also called mitochondrial autophagy. At present, the researches on autophagy is mainly focused on macro autophagy.

At present, it is generally believed that autophagy is a mechanism of protection and stress regulation, and can be divided into background autophagy and induced autophagy according to the occurrence of the situation. Low-level background autophagy persists in eukaryotic cells and is considered to be very important for the maintenance of cell homeostasis, and is a self-protection mechanism of cells. A high level of induced autophagy is a stress mechanism of cells. Various factors include nutritionally deficient cell starvation, physical factors such as radiation, and various chemical inducers such as the inhibitor rapamycin, conventional anticancer drugs etoposide, mTOR-independent inducer Trehalose, etc. They all lead to a significant increase in cellular autophagy.

1.3.4 Occurrence and Detection of Cell Autophagy

1.3.4.1 The Process of Autophagy

The occurrence of autophagy is a dynamic process. In this process, the autophagosomes from the rough endoplasmic reticulum, Golgi apparatus, etc. fall off to form a

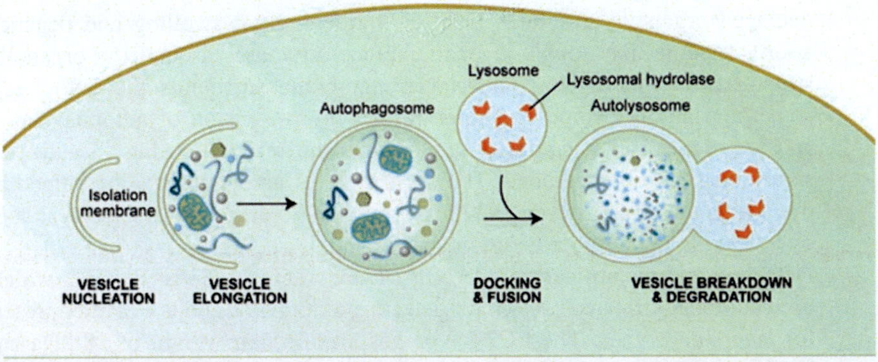

Fig. 1.4 The process of autophagy [60]

cup-shaped partition membrane firstly, which surround the substance to be degraded, and then the diaphragm gradually extend to form autophagosomes. Autophagosomes are transported to the lysosomes through the cytoskeleton microtubule system, and then fuse with lysosomes to form autolysosomes. Next, the encapsulated components would be degraded, and the autophagosome membranes would be shed and recycled (Fig. 1.4) [60]. More than 35 autophagy-related genes (called ATG genes) have been found to participate in the process of autophagy regulation [54, 61].

1.3.4.2 Detection Method of Autophagy

In 2008, the first edition of "guidelines for the use and interpretation of assays for monitoring autophagy" in higher eukaryotes" was published in *Autophagy* by 214 researchers in the field of autophagy, which established a series of scientific standard methods for detecting autophagy in higher eukaryotes [61]. With the continuous development of autophagy research, a large number of scientists have entered the field, and the knowledge base and detection technology of autophagy have been updated. Therefore, in April 2012, Daniel J. Klionsky led nearly a thousand scientists in the field of autophagy to organize and publish the second edition of the "guidelines for the use and interpretation of assays for monitoring autophagy" on the basis of the first edition in 2008, to further optimize and refine the identification criteria and research methods in the autophagy discovery and occurrence for different model organisms and different systems process [62]. The following are some common and basic detection methods mentioned in the article:

1. Transmission electron microscope: Observing the morphology of cell ultrastructure using a transmission electron microscope is the gold standard for autophagy detection. When autophagy occurs, the maturation from autophagosomes to autolysosomes is a dynamic and continuous process [63]. Different morphological structures can be observed under transmission electron microscopy:

damaged organelle structures (such as mitochondrial swelling and degeneration), vacuole-like double-layer membrane structures around the organelle, structures in which the double-layer membrane surrounds the autophagosome, and autolysosome structures formed by the fusion of autophagosome and lysosome, etc. It can also be seen that the residual bodies in autolysosomes unable to be degraded. The occurrence of autophagy can be judged by observing autophagosomes, and the degree of autophagy can also be evaluated by observing the number of autophagosomes.

2. Detection and quantification of *ATG8* / LC3: *ATG8* / LC3 is the most widely used autophagy-related protein for detecting autophagy, and is a marker protein for autophagosomes. The LC3-I type has a molecular weight of 18 kDa and is distributed in the cytoplasm; the LC3-II type is 16 kDa and is mainly distributed on the surface and inside of autophagosomes. When autophagy occurs, autophagosome formation increases, and LC3-I is specifically cleaved into LC3-II and then transferred to the autophagosomes. The overall level of the protein does not change, and the occurrence of autophagy can usually be determined by detecting the conversion of LC3-I to LC3-II by immunoblotting.
3. Florescence microscope observation: When autophagy occurs, the aggregation of LC3 protein on autophagosomes can be seen in the cell. The LC3 antibody can be used for immunofluorescence to observe the accumulation of endogenous LC3 or transfecting the exogenous fluorescent protein GFP-LC3 protein, and observe the number of green dot-like aggregates produced in the cell to judge and evaluate the occurrence of autophagy [64]. Observe LC3 protein and acidic organelles colocalizations: Stain the organelles with acid dyes orange or MDC, ang then observe the exogenous GFP-LC3 green light spot-like aggregation and co-localizations of these stained markers to determine the occurrence of autophagy.
4. Dynamic detection of Autophagy flux: Induction of autophagy is not just a simple process of increase in LC3-II or autophagosome formation. It is more important to observe the dynamic multipolarity of the entire autophagy system including lysosomes. The increase of intracellular autophagosomes may be caused by the upstream autophagy process being stimulated, or by the reduction of downstream autophagosome degradation. Therefore, in conjunction with the detection of the above methods, the changes of substrate proteins, such as p62 / SASTM1, LC3-II, NBR1, BHMT, NeoR-GFP, etc., must also be detected [65–67].
5. Detection of long-term protein degradation: This is a very effective detection method that can achieve the purpose of quantifying the degree of autophagy. Proteins in cells are usually labeled with radioactive isotopes, such as [^{14}C] -leucine, [^{14}C] valine, or [^{35}S] methionine, and the cells are cultured for a long period of time so that they can be specifically labeled long-term proteins, while short-term proteins have been degraded by the proteasome pathway. After the occurrence of autophagy, the amount of acid-soluble radioisotopes released in the cell culture medium can be used to quantify the autophagy-degraded proteins in the cell [68–70].

1.4 Physiological and Pathological Significance of Autophagy

In fact, autophagy is a kind of "double-edged sword" for cells. Low levels of autophagy persist in cells, and are used by cells to remove damaged or no longer needed organelles, proteins, and extracellular pathogens to maintain cell homeostasis, which is a necessary Protection mechanism for cell survive. On the other hand, when the environment changes, autophagy can induce cell damage and lead to cell death. Cell autophagy is closely related to important physiological processes and has important physiological and pathological effects on the body, and is related to the occurrence, development and treatment of variety of diseases.

1.4.1 Autophagy and Cancer

A large number of studies have shown that cancer, a major disease that seriously harms human health, have an intricate relationship with autophagy. Autophagy is closely related to the occurrence, development and treatment of cancer. It has different or even opposite effect on different stages of tumorigenesis.

1.4.1.1 Autophagy and Cancer Occurrence and Development

Basic autophagy that occurs in normal cells can maintain chromosome stability and prevent cell necrosis and inflammatory infiltration, thereby reducing the chance of canceration in normal cells [71]. Studies have shown that histone deacetylases in yeast cells can repair DNA damage through autophagy and thus stabilize the structure and function of chromosomes. In contrast, the loss of autophagy usually leads to the abnormal accumulations of damaged organelles and macromolecular components within the cell, which induces cellular damage effects such as oxidative stress, DNA damage and chromosomal instability, and eventually leads to canceration of cells [72]. The most typical example is the mouse with the Beclin 1 gene (the first autophagy-related gene identified in mammals, a family member of Bcl-2) knocked out, whose tumor incidence are significantly higher and occurrence time earlier than wild type mice [73]. At the same time, Beclin 1 single allele deletion is closely related to breast cancer, ovarian cancer and many other cancers. Subsequent researches had shown that the defect in autophigic function caused by the deletion of a single allele of Beclin 1 may be the important reason for the increased risk of DNA double-strand breaks in cells, abnormal gene replication and aneuploidy in cells [74]. In addition, factors like abnormal aggregation of p62, endoplasmic reticulum molecular chaperones, damaged organelles, and oxidative free radicals in phagocytic tumor cells, have increased the tumorigenic capacity of tumor cells to a certain extent [75]. In the early stage of cancer, inhibition of autophagy causes continuous growth of pre-cancer cells, indicating that autophagy has the effect of suppressing tumors. These results all show the relationship between inhibition of autophagy and tumorigenesis [76]. Meanwhile, autophagy can also suppress tumorigenesis by reducing the mutation rate by eliminating damaged organelles [77].

On the other hand, autophagy plays a positive role in promoting the development after tumor formation. Tumor cells that proliferate at a high rate are often in a microenvironment of hypoxia and nutritional deficiencies after tumor formation. At this time, it is through the increase of autophagic level that tumor cells can provide sufficient energy for their survival and growth. Studies have shown that in bladder cancer and lung cancer cells, the expression of proto-oncogenes H-ras and K-ras can up-regulate the level of self-proliferation, thereby promoting the growth of cancer cells under stress [78]. In pancreatic tumor cells and primary pancreatic cancer cells with a high level of autophagy, the inhibition of autophagy can reduce the level of intracellular mitochondrial oxidative phosphorylation, increase free radicals and DNA breakage, and then kill cells [79]. In addition, under metabolic pressure, autophagy promotes tumor development more prominently. In vitro and in vivo experiments have shown that in the ischemia or hypoxia microenvironment, tumor cell viability with autophagy deficiency is significantly low [80].

1.4.1.2 Autophagy and Cancer Treatment

Autophagy has a protective effect on tumor cells, so using autophagy inhibitors to reduce autophagic levels can enhance the sensitivity of tumor cells to therapeutic drugs (such as proteasome inhibitors, etc.). In vitro and in vivo studies have shown that the combined use of the proteasome inhibitor bortezomib and the autophagic inhibitor chlorine can increase tumor cell mortality [81]. At present, there are many studies on using autophagy inhibitors to enhance the efficacy of tumor chemoradiotherapy. Among them, the combination of hydroxychloroquine with clinical drugs to inhibit autophagy for the treatment of malignant tumors has entered clinical trials in the United States and Europe, and is specifically used to treat various of solid tumors including lung cancer, breast cancer, prostate cancer, pancreatic cancer, kidney cancer, brain cancer, and central nervous system cancer. How to achieve more effective tumor treatment through more effective regulation of autophagy and coordination with conventional methods is bound to become a new research hotspot [82]. In addition, many clinically used anti-tumor drugs can trigger autophagy, including histone deacetylase inhibitors, anti-angiogenesis inhibitors, mTOR inhibitors, BH3 domain analogs, and glycolytic enzyme inhibitors. A notable example is 2-DG, which is in the Phase 1 clinical research phase (used to treat many cancers including prostate cancer). 2-DG is a typical glycolysis inhibitor that can effectively reduce the level p62 protein, and thus induces cell autophagy [83]. Autophagy plays a key role in the treatment of these drugs. In addition, autophagy may be an important cause of drug resistance in cancer cells. Increased autophagy levels in cancer cells help cancer cells to resist the stress of drugs, which in turn leads to the development of drug resistance during cancer treatment. Therefore, how to counteract the drug resistance of tumor cells by reducing the autophagy level of tumor cells and then improve the killing effect of chemotherapy drugs is also a new strategy for tumor treatment.

1.4.2 Autophagy and Neurodegenerative Diseases

An important function of autophagy is to remove protein deposits in cells and prevent the occurrence of neurodegenerative diseases. Studies have found that the accumulation of autophagosomes in nerve cells in brain diseases [84], indicating that autophagy plays an important role in maintaining the stability of the inner environment of nerve cells and controlling protein quality [85]. There must be some mechanism in nerve cells to detect these functionally deficient substances and to degrade pathogen protein aggregates (such as α-synuclein, PINK1-PAEKIN, synphilin-1, mutant tau protein or Huntington's protein) by autophagy, before these substances become toxic to the nerve cells themselves, while the aggregates of p38 and myosin cannot be recognized by it [86, 87]. ATG5 gene knockout causes autophagy to not work properly, and a large number of abnormal aggregation of proteins and inclusions appears in mouse neurons, which cause neurodegenerative diseases [88]. Similar results were also found in CNG-specific ATG7 knockout mice, where a large number of neurons were lost in the mouse's brain and cerebellar cortex, resulting in behavioral defects in mice [89]. In addition, the occurrence of neurodegenerative diseases can lead to impaired autophagic function. Lee J.H. et al. found that blastocysts lacking the PS1 protein associated with Alzheimer's disease. Lysosomal acidification and cathepsin activation are selectively destroyed, resulting in degradation of substrate proteins and obstruction of autologous carcass clearance [85]. Therefore, increasing autophagy levels by inducing autophagy may be a good way to treat these neurodegenerative diseases.

1.4.3 Autophagy and Development

Autophagy plays a key role in both individual development and tissue and organ differentiation. Autophagy is initiated during the fertilization stage and is up-regulated during early embryonic development. Fertilized eggs lacking the ATG5 gene decrease in protein synthesis speed and stop development at the 4–8 cell stage [90]. Maternal material such as the mitochondrial genome in the fertilized egg is eliminated by a autophagic pathway so that it does not remain in the embryo [91]. During embryonic development, the organism clears dead cells by autophagy, while dead cells in ATG5 knockout mouse embryos cannot be removed normally. After birth, an individual consumes its own proteins to produce nutrients to help the organism survive the hunger period [92]. After birth, the ATG5-deficient mice have a significantly shorter survival period than wild-type mice [93].

1.4.4 *Autophagy and Infection, Immune and Inflammatory Response*

Autophagy plays a vital role in pathogen infection and immune system response. The hepatitis virus antigen EBNA1 needs to be processed and presented by the autolysosomal pathway, and then recognized by T cells [94]. At the same time, autophagy is necessary for organisms to build self-tolerant T cell banks and it plays an important, sometimes even distinct, role in resisting pathogen infection [95]. Autophagy plays a natural immune response role in against intracellular bacterial and viral infections. On the one hand, autophagy promotes the body's elimination of intracellular pathogens and the processing and presentation of antigens. Cells can recognize single-stranded RNA viruses through autophagy and clear intracellular mycobacterium tuberculosis and streptococcus infections [96–98]. Nucleic acid antigens of intracellular viruses can be presented to MHC-II-like molecules through autophagy. Antigen-presenting cells can process antigens by autophagy processing and present antigen peptides in the form of MHC-I or MHC-II-type molecules to specific T lymphocytes [94, 99]. On the other hand, intracellular infection pathogens evade the body's autophagy through some mechanisms. Under normal circumstances, the invading bacteria are endocytosed by the host cells, then come out of the endocytic vesicles and enter into the cytoplasm. The autophagosomes immediately wrap them and transport them to the lysosomes for degradation. The host can clear many bacteria that invade the host, such as Listeria, Mycobacterium tuberculosis, Shigella flexneri, Streptococcus, Turabacterium, and so on [100]. The IcsB protein secreted by the dysentery bacillus through the type III secretion system inhibits host cell autophagy, thereby escaping host cell degradation [101]. Some viruses use autophagosomes as sites for virus replication to help them infect cells [102]. Recent studies showed that not all autophagy promote clearance of intracellular pathogens in host cells. In some cases, autophagy is conducive to the survival of intracellular pathogens. For example, during some bacterial infections, such as A. odophyra and Legionella pneumophila, they can replicate in autosome-like vesicles. Also in some cases, autophagy is not a matter of removing virus particles, but of helping the virus replicate [103]. Some pathogens (such as rhinovirus and poliovirus) hide themselves in autophagosomes and induce cells to undergo autophagy after infecting the host. At this time, autophagosomes provide a membrane structure for viral RNA replication, which in turn helps offspring virus particles to be released from infected cells [104–106]. Therefore, in view of the role of cellular autophagy in pathogen infection and immune system response, rational regulation will play an important role in the treatment and intervention of pathogen infection.

1.4.5 Autophagy and Metabolism

Autophagy plays an indispensable role in the metabolism of organisms. The flux of L-glutamic acid regulates the mTOR pathway and protein translation, and also coordinates cell growth and proliferation through autophagy [107]. Under stress, autophagy plays an important role in cell lipid metabolism. The decreased level of autophagy in mouse liver cells will greatly increase the triglyceride content in the lipid droplets [108]. "Exercise is good for health". Its molecular mechanism is directly related to autophagy. Under the state of vigorous exercise, the interaction between Bcl-2 and Beclin 1 is weakened, which leads to a significant increase in autophagy and triggers glucose metabolism changes to meet the demands of strenuous exercise. The exercise capacity of mice with autophagy is significantly reduced [109].

1.4.6 Autophagy and Aging

Autophagy is closely related to individual aging. Morphological and enzymatic changes in the lysosomal system are present in almost all aging tissues. With the increase of age, the effect of autophagy begins to weaken, resulting in a decrease in the cell's ability to defend itself and adapt to the external environment. Damaged organelles and other cellular structures and reactive oxygen species such as a large number of oxygen free radicals cannot be removed in a timely and effective manner. Studies have shown that the additive effect of mitochondrial inactivation and autophagy disorders is one of the important causes of aging in organisms [110]. Autophagy can prevent the aging of neurons by degrading misfolded proteins and damaged organelles [111]. In addition, the developmental stage of worms and life cycles are all regulated by autophagy. Autophagy related genes such as ATG1, ATG6, ATG7, ATG8 and ATG10 play critical roles in above process, and provides a genetic basis for the correlation between autophagy and aging [112, 113].

1.5 Autophagy is a Special Biological Effect Possessed by Many Nanomaterials

1.5.1 Research Status and Development Trend of Nanomaterial-Regulating Autophagy

With the continuous development of the application of nanomaterials in the field of biomedicine, the research on the biological effects of nanomaterials has also become a research focus in this field. As an important cellular biological effect of nanomaterials, autophagy has been paid more and more attention worldwide. Studies have shown that many inorganic and organic nanomaterials can trigger autophagy in a variety of cells, Therefore, increasing autophagy levels may be a universal response of cells to nanomaterials.

1.5.1.1 Study on Inorganic Nanomaterial-Regulating Autophagy

In 2005, Wen Longping's group at the University of Science and Technology of China reported for the first time that inorganic nanomaterial Nd_2O_3 can trigger cell autophagy in human large cell lung cancer cell line NCI-H460 [114]. In 2006, Seleverstorv et al. compared the cytotoxicity and intracellular processes of two sizes of quantum dots in human bone marrow mesenchymal stem cells, and found for the first time that the autophagy induced by nanomaterials was scale-dependent [115]. Subsequent research by Stem et al. also showed that two quantum dots of equal size but different chemical composition can both induce autophagy in pig kidney cells, and suggested that inducing autophagy may be a universal cellular biological effect of nanomaterials [116]. In 2009, Zhang et al. found that fullerene C_{60} induced autophagy in cells through ROS, and C $_{60}$ could increase the sensitivity of various cancer cells to chemotherapy drugs [117]. In the same year, Yu et al. reported that rare earth metal oxide nanocrystals with different particle sizes, including light rare earth (such as Sm, Eu) metal oxide nanocrystals and heavy rare earth (such as Gd, Tb) metal oxide nanocrystals can cause autophagy, and proposed that triggering autophagy is the commonality of nano rare earth metal oxides [118]. This work has received extensive attention from the research field of autophagy. At the invitation of Professor Klinosky, the editor-in-chief of *Autophagy*, professor Wen Longping wrote a review article in this field and published it in the 2010 issue of the journal [119]. Liang Xingjie's research team found that gold nanoparticles entered cells through endocytosis and aggregate in lysosomes, which in turn changed the acidic environment of lysosomes, leading to inactivation of lysosomal enzymes and impaired lysosomal degradation. Gold nanoparticles destroyed the complete autophagy flux and a large number of autophagosomes being accumulated in cytoplasm [125]. So far, research on inorganic nanomaterials with the ability to regulate autophagy has mainly included carbon nanomaterials (such as carbon nanotubes [120] and fullerene and their derivatives, etc. [117, 121, 122]), rare earth metal oxide nanomaterials [118, 123], and magnetic nanomaterials (such as Fe_3O_4 [124], MnO, etc.), quantum dots, and metal nanoparticles (such as gold nanoparticles [125, 126], silver nanoparticles, etc.).

1.5.1.2 Study on Organic Nanomaterial-Regulating Autophagy

Organic nanomaterials have unique advantages in terms of drug transport and degradability. At present, the representative work of organic nanomaterials to regulate auophagy mainly includes: Li et al. discovered that PAMAM dendrimers with a particle size below 6 nm can induce autophagic cell death by inhibiting the AKT-TSC2-mTOR signaling pathway in 2009 [127]; Man et al. found that cationic liposomes with a particle size of about 100 nm can induce mTOR-independent integrity autophagy in 2010 [128]. In addition, ethyl acrylate [129] and PEG-PE nanomaterials have also been shown to have the ability to induce autophagy.

1.5.2 Nanomaterial-Regulating Autophagy Provides Significant Opportunities for Tumor Treatment

Autophagy is a biological process closely related to cell death. During cell death, a sharp increase in autophagic vesicles is usually observed [130]. There is also a strong link between autophagy and two common forms of cell death (apoptosis and cell necrosis). Studies have shown that autophagy is an important cause of apoptosis and necrosis of cells [131]. Nanomaterials, as a new type of autophagy inducer, usually promote autophagy and cause cell death [127]. This autophagy-mediated cytotoxicity provides a theoretical basis for the application of nanomaterials in cancer chemotherapy and radiotherapy. Although nanomaterials have been used in the diagnosis and treatment of cancer, most of the research at this stage is still based on the structure and function of nanomaterials. The research combining the cellular effects of nanomaterials, especially the autophagy effect with cancer diagnosis and treatment is still preliminary.

In terms of cancer diagnosis, the diagnosis of cancer mainly depends on medical imaging. At present, magnetic resonance imaging contrast agents and other imaging and tracer agents with clinical application prospects have been shown to induce cell autophagy, such as the current research hotspot of nanometer magnetic resonance imaging contrast agents, superparamagnetic MnO Nanomaterials, iron oxide nanoparticles and QDs [116] for fluorescent imaging. At present, the main research directions of magnetic contrast agents are to improve the contrast ability of materials, and to optimize their body distribution and metabolism through surface modification. Since autophagy is an important aspect of the biological effects of contrast nanomaterials, related research has also received increasing attention. Khan et al. found that Fe_3O_4 caused mitochondrial damage and induced autophagy in cancer cells by producing ROS [124]. MnO can induce cancer cells to undergo autophagy, enhance chemotherapy killing, and achieve a good integration of diagnosis and treatment [132]. These studies provide two directions for clinical diagnosis and treatment of cancer: when nanomaterials are used as imaging reagents only for cancer diagnosis, the safety of such imaging reagents can be improved by avoiding autophagy, and it is possible to improve the imaging noise ratio; and when the nanomaterial is used not only to diagnose the location of the lesion, but also to produce a therapeutic effect, we can use the cytotoxicity induced by the autophagy to kill cancer cells.

In cancer chemotherapy, the autophagy effect of nanomaterials can promote cancer chemotherapy. Previous studies have shown that nano-gold particles [133] and nano-gold (iron) particles [126] may perform effective chemotherapy through autophagy [130], while rare earth metal potentiated fullerene derivatives are used in combination with cisplatin to greatly enhance its effect of chemotherapy in resistant cancer cells [3]. Cell autophagy induced by nanomaterials plays a synergistic role in killing cancer cells in the chemosensitization of doxorubicin and cisplatin. Studies have found that nanofullerene C60 and its derivative C60 (Nd) can induce cancer cells to develop autophagy, and can combine the chemotherapy drugs doxorubicin and cisplatin in cervical cancer cells to exert anti-cancer effects. Further research shows that the autophagy effect of nanofullerene and its derivatives can also improve sensitivity to doxorubicin and enhance the effect of chemotherapy in MCF-7-MDR [121]. These research results provide a theoretical basis for the nanomaterials to be used for chemotherapy of cancer cells through the autophagy effect.

In terms of cancer immunotherapy, cell autophagy induced by inorganic nanomaterials is involved in cancer cell immunotherapy. The study found that α-Al_2O_3 nanoparticles in the dendritic cells passed antigens to autophagosomes to help cells complete antigen presentation by inducing autophagy. This nanomaterial is expected to develop into a therapeutic cancer vaccine [134].

In terms of cancer radiotherapy, non-nanomaterial drugs can be widely studied as synergists for cancer radiotherapy. For example, chloroquine, as a drug that antagonizes autophagy, can increase the sensitivity of cancer cells to radiotherapy. It has entered the phase I clinical stage in the Netherlands and is used to treat stage IV small cell lung cancer, which can effectively improve the overall survival rate of small cell lung cancer patients. Arsenic trioxide has also been shown to increase the sensitivity of cancer cells to radiotherapy by inducing cell autophagy [135]. In addition, nano-silver particles can induce autophagy in cervical cancer cells, and increase the sensitivity of cancer cells to radiotherapy, thereby enhancing the effect of radiotherapy, making nanomaterials a possible radiotherapy synergist.

The autophagy induced by nanomaterials can effectively promote the killing of tumor cells by nanomaterials and achieve the purpose of cancer treatment. However, for normal cells, the autophagy effect induced by nanomaterials has shown serious of biological effect. The issue of security has caused people to think.

1.5.3 Biosafety Issues Caused by the Autophagy Effect of Nanomaterials

Low levels of basic autophagy in cells help maintain homeostasis, and under many physical, chemical, and biological pathological factors, autophagy levels in cells often increase significantly. This "inducible" autophagy may be functional (normal or complete autophagy) or non-functional (abnormal, excessive, or incomplete autophagy, that is, the process of autophagy can only advance to a certain stage. As a result, the contents of autophagy cannot be degraded by lysosomes). Excessive and abnormal autophagy can negatively affect cells and even cause cell death. In recent years, a number of research results have shown that many nanomaterials can cause cell autophagy. The nanomaterials that have been reported to cause autophagy are rare earth oxides, quantum dots, fullerenes, carbon nanotubes, PAMAM dendrimers, gold particles, TiO_2, etc. However, most of these nanomaterials induce pro-death autophagy. This cytotoxic effect may be due to that nanomaterials promote the occurrence of cell autophagy while destroying the integrity of autophagy. Recent studies have found that autophagosomes would be broken during autophagy induced by rare earth oxide nanocrystals. This phenomenon suggested that after inorganic nanomaterials enter autophagosomes and fuse with lysosomes, the enzymes in the lysosome cannot degrade the nanomaterials, further leading to release of a variety of acid hydrolytic enzymes from lytic lysosomes, which eventually triggers disturbances in the normal physiological processes inside the cell, and leads to cell death.

Nanomaterial-caused autophagy has deepened people's concerns about the safety of nanomaterials. It is well known that air pollutants contain a large number of nanoparticles, and epidemiological studies have shown that compared with areas with low air pollution, the incidence of Parkinson's disease, Alzheimer's disease and other neurological diseases in areas with high air pollutions significantly increased in Mexico City. The level of autophagy has significantly changed in the brain of patients with neurodegenerative diseases. This study further suggests that the autophagy effect is an important factor affecting the biological safety of nanomaterials. The direct evidence comes from the study of PAMAM dendrimers. PAMAM dendrimers have broad application prospects in many aspects such as drug carriers, gene carriers, nanocomposites, and nanoreactors. Jiang Chengyu's group reported that PAMAM, a nanomaterial used as a drug carrier in medicine, causes cell death by inducing autophagic signaling pathways, and inhibiting autophagy can effectively reduce lung injury. This research results revealed the molecular mechanism of nanomaterials causing lung injury through autophagy for the first time, and provided a theoretical basis for preventing the toxic and side effects caused by the autophagy of nanomaterials, which has attracted widespread attention in the academic community and media [127].

In addition, recent studies have shown that the biosafety of nanomaterials is closely related to its induction of cell autophagy. In addition to the work of the aforementioned PAMAM, the water-soluble fullerene derivative $C_{60}OH_x$ leads to impaired mitochondrial function of renal cells and increased levels of cellular autophagy, which triggers cytotoxicity [136]. Fulleronol toxicity is due to reduced intracellular mitochondrial oxidative phosphorylation and increased autophagosomes by nanomaterials [137]. Polyalkyl sulfonated fullerene FC4S caused lysosomal overload in rat kidney cells, and decreased lysosomal degradation capacity with increased intracellular autophagosomes, eventually causing renal disease [138]. The 50 nm carbon black nanoparticles in cigarettes affected the conversion of autophagosomes to lysosomes in human lung macrophages, and increased autophagosomes and intracellular aggregates, resulting in damage to mitochondria and weakened oxidative stress capabilities, finally causing lung disease [139].

In summary, in the development and application of nanomaterials, the phenomenon of autophagy caused by nanomaterials cannot be ignored as a safety issue. Some nanomaterials that enter the human body through daily circulation such as water circulation and atmospheric circulation need to avoid and eliminate the cell autophagic effect. When the various characteristics of nanomaterials are used as drug carriers or diagnostic reagents directly in clinical applications, autophagy also need to be circumvented to improve their safety. Even for nanomaterials that use the autophagy effect as a treatment method, the autophagy in non-action sites and non-target cells is also a manifestation of the side effects of nanomaterials, which should be avoided. Therefore, it is important to further explore research methods that regulate cell autophagy to improve the biological safety of nanomaterials. The relationship between the autophagy effect of nanomaterials and biosafety suggests that the biosafety of nanomaterials can be improved from three aspects to meet the requirements of biomedical applications: ① The composition, electronic structure, surface bonding substances, surface coverings, etc. of nanomaterials can be improved by changing the characteristics of nanomaterials, further changing their abnormal adsorption capabilities, chemical reaction capabilities, dispersion and agglomeration capabilities, etc., thereby affecting their autophagy capabilities and thus improving biosafety; ② By regulating the ability and degree of nanomaterials to induce autophagy, such as the use of autophagy regulators to change the ability and degree of nanomaterials, and thus improving the biological safety of nanomaterials; ③Modify nanomaterials by targeting groups to enable autophagy to occur in specific cells, tissues and organs. And circumvent its ability to induce autophagy in non-targeted organs, thereby improving the biological safety of nanomaterials.

1.6 Modification Methods Using Peptide Enables Nanomaterials to Effectively Regulate Autophagy

The surface characteristics of nanomaterials have a great relationship with the ability of nanomaterials to induce autophagy and biological safety. It becomes important issues that how to regulate the surface properties of nanomaterials to regulate the autophagy activity of nanomaterials induced by cells, how to improve the biological safety of materials, and how to improve biocompatibility of materials in tumor diagnosis and treatment and the ability to regulate cell autophagy for disease diagnosis and cancer treatment, which need to be addressed urgently [140].

1.6.1 Modification and Surface Modification of Nanomaterials

Unlike ordinary materials, nanomaterials have a small particle size with a large specific surface area and a high surface energy, so nanomaterials are extremely prone to agglomeration and dispersion in applications. When they are used in biomedical detection and treatment, the final fate of nanoparticles in the body is determined by the interaction between the nanomaterial and the surrounding environment in the body. And this interaction depends on the size and surface properties of the nanomaterial. In order to ensure that nanometer materials always exist in the size of nanometer level during the preparation and application, the surface modification of nanometer materials has become a key issue for nanometer materials research.

In the 1990s, the International Nanometer Conference first proposed the concept of surface modification. With the rapid development of nanotechnology in recent years, the surface modification technology of nanomaterials has also been widely studied. Surface modification technology of nanomaterials mainly refers to the use of physical or chemical methods to change the structure and state of the surface of nanomaterials, to give new properties to nanomaterials, and to improve their physical properties (such as particle size, charge, etc.), which is convenient for a wider range of application. According to the interaction between nanomaterials and modifiers, surface modification technologies mainly include physical modification methods (such as surfactant method, surface deposition coating, etc.) and chemical modification methods (such as coupling agent method, alcohol esters chemical reaction method, acid flower reaction method, surface graft modification method, etc.). When nanomaterials are used in biomedical applications, the most widely used method is to modify the surface atoms of nanomaterials with the coating of modified molecules (mainly organic substances such as chemical molecules and biomolecules) in order to change the surface structure and state of nanomaterials.

In recent years, scholars from various countries have done a lot of work on surface modification of nanoparticles to optimize the performance of nanomaterials and promote biomedical applications, which have made great progress. At present, dozens of surface-modifying molecules have been developed for nanomaterials, among which the most widely used are chemical molecules (surfactants such as chitosan, silicon, and PEG, PVP, PEI, SDS, PLL, TTAB, fatty acids, etc.) and biomolecules (such as peptides, oligonucleotides, proteins, etc.). Chitosan-coated magnetite nanoparticles can effectively improve the dispersibility of iron oxide and can be used as an adsorbent for the extraction and analysis of trace amounts of environmental water samples [141]. PEG is a polymer of ethylene oxide and can be coated with superparamagnetic iron oxide to form monodispersed iron oxide nanoparticles for MRI imaging [142]. TTEB is a quaternary ammonium compound. It is a cationic surfactant, and is widely used in the synthesis and dispersion of nano silver, which has a good bactericidal effect [143]. Peptides are widely used in the self-assembly and crystallization of nanomaterials. As early as 2000, studies have shown that peptides can guide the self-assembly of semiconductor materials [144] and can also direct the crystallization of nano-gold particles, reducing gold chloride to nano-gold crystals [145]. Other studies have shown that oligonucleotide-functionalized gold nanoparticles are more easily taken up and internalized by cell membranes [146].

Due to their extremely high surface energy and surface activity, nanomaterials are extremely prone to adhesion between nanomaterials and between nanomaterials and the environment. This kind of non-specific adhesion severely restricts the application of nanomaterials. Especially when used in biomedicine, nanomaterials are prone to non-specific adhesion to the cell surface and the inner surface of blood vessels, which greatly reduces the effect of nanomaterials. Therefore, nanomaterials must be surface modified before they can be effectively applied. Among the many modification methods, the worthiest of mention is the non-covalent modification method using high-affinity peptides, which can be modified without changing the properties of the nanomaterial itself. It is widely used in the research of tissue and cell engineering of nanomaterials.

1.6.2 Regulation of Peptides and Proteins on Nanomaterials

In order to make nanomaterials have good biocompatibility and cell affinity, the surface modification of materials is particularly important. There are many modified molecules used to achieve surface modification of nanomaterials. But considering compatibility with the biological environment, peptides are the most common and most convenient one. Polypeptides used as surface modifiers for nanomaterials have several outstanding points: firstly, the peptide is composed of amino acids, which is closest to the biological environment, and is the best biocompatible surface modifier [147]; secondly, peptides have versatility. Peptides of different sequences have different biological characteristics. For example, RGD can specifically bind to the cell surface receptor, integrin, and integrin is highly expressed on a variety of tumor surfaces and neovascular endothelial cells, which is important for tumor angiogenesis. So the nano-materials connected to RGD successfully achieved the targeting of tumor cells [148]; In addition, peptide synthesis is simple and highly controllable. With the continuous development of science and technology, the synthesis of peptides has become industrialized, and the purity can reach 99.99%. A variety of peptides of different lengths and sequences can be synthesized for the surface of nanomaterials according to the characteristics and application requirements of nanomaterials modification. In addition, peptides or proteins also have the effect of changing the interaction between nanomaterials and cells and regulating the toxicity of nanomaterials. Studies have shown that peptides and proteins can regulate cell uptake capacity and toxicity in cells of single-walled carbon nanotubes and nano-amorphous silica [147, 149]. Therefore, we envisage whether it is also possible to use surface modification methods of specific peptides to achieve the regulation of autophagy ability of nano-induced cells.

1.6.3 *Screening of Peptides that Specifically Bind to Nanomaterials Using Phage Display Technology*

Using phage display technology to select peptides that specifically bind to nanomaterials, only simple mixing is required to make the nanomaterials bind to specific peptides with high affinity. This binding is different from nonspecific adhesion with lower affinity, and also different from the covalent bonding due to the violent chemical reaction, which can achieve high unbalanced bonding only under mild conditions.

1.6.3.1 What is Phage Display Technology

Phage display technology refers to the biotechnology that a polypeptide or protein coding gene, or a gene fragment of interest, clones into an appropriate position of a phage coat protein structural gene. When the reading frame is correct and does not affect the normal function of other coat proteins, the foreign polypeptide or protein is fused and expressed. Finally, the fusion protein is displayed on the surface of the phage with the reassembly of the progeny phage.

1.6.3.2 Origin of Phage Display Technology

In 1985, the genome of the filamentous bacteriophage was transformed by genetic engineering for the first time. The exogenous gene was fused to the end of the filamentous bacteriophage coat protein gene so that the polypeptide encoded by the gene of interest was displayed on the surface of the bacteriophage as a fusion protein, thereby creating the bacteriophage Show no technology [150].

1.6.3.3 Structure of Filamentous Phage

The filamentous phage has a fixed diameter of about 6.5 nm, and its length is determined by the size of its genome. Phage particles are composed of 5 capsid proteins, namely P III, P VI, P VIII, and P IX. These 5 capsid proteins can be used for the display of foreign proteins (Fig. 1.5). Among them, P III and p VIII proteins are currently the most extensive capsid protein. The P III protein has 406 amino residues and is the most commonly used capsid protein in phage display. Located at the tail of the phage particle, there are 3–5 copies that can be inserted into two sites of the foreign protein, and the N-terminus is the fusion site of the foreign protein. P VIII protein is the main surface protein of phage particles. The hollow tubular structure surrounding single-stranded DNA is composed of thousands of 50 amino acid residues of P VIII protein. Because P VIII protein is small, when it carries peptides that are too large, will form steric hindrance, which will affect the assembly of the shell and lose infectivity; P III protein has no strict limit on the size of the exogenous

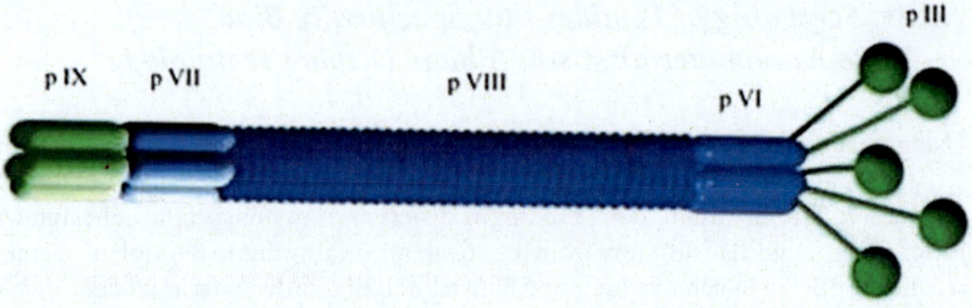

Fig. 1.5 The structure of filamentous phage

polypeptide or protein displayed, so it is most commonly used for phage display. The filamentous phages used in the phage display technique herein are all M13 phages.

1.6.3.4 Features of Phage Display Technology

The main feature of this technology is that the genotype and phenotype of a specific molecule are unified in the same virus particle [151], and the exogenous polypeptide or protein to be expressed is displayed as a fusion protein on the surface of a phage, which can be screened by affinity enrichment. Bacteriophages expressing specific peptides or proteins lay the foundation for selecting target peptides or proteins in the future.

1.6.3.5 Selection of Phage Display Technology

There are two classical screening methods: solid phase screening and liquid phase screening. Solid-phase screening refers to directly coating the target molecule on a solid-phase medium, such as an ELISA plate, an immune test tube, and an affinity chromatography column. And then adding the phage to be screened, washing away non-affinity phage, and recovering high affinity bacteriophage and further amplified. Liquid phase screening is a method for screening phages by using target molecules linked to biotin groups and immobilizing them on paramagnetic beads coated with saturated streptavidin [152, 153]. The elution process of the two methods is the same: they are coated with the target molecule firstly, then non-specific sites are blocked with BSA, and the phage peptide library is added to allow the peptide library and the target molecule to incubate for a certain period of time before washing the unbound free phages. Finally get the specifically binding phage with the target molecule under the elution of a competitive receptor or an acid. After the eluted phage infects the host cells, it is propagated and amplified, and the next round of selection is performed. After 3–5 rounds of "adsorption-elution-amplification", phages that specifically bind to the target molecule are highly enriched. The resulting phage can be used for further screening and enrichment of target phages with binding characteristics.

With the development of phage display technology, phage screening methods have been further studied. According to the different target molecules selected, screening methods for intact cells, tissues and organs have been developed. As early as 1996, Pasqualini et al. injected phage intravenously into mice and screened short peptides that bind to the blood vessels of mouse brain and kidney tissues. Since then, phage display technology has developed rapidly in the biomedical field [154]. In 1998, D Rajotte screened two short peptides that specifically bind to tumor neovascularization with high affinity, with sequences of RGD and NGD, respectively [155]. A large amount of research work was subsequently developed, and a variety of cells and cell line-specific peptides were selected using the random peptide library of the phytoplasma. The cell lines included highly metastatic gastric cancer cells [156], multiple ovarian cancer cells [157], colon adenocarcinoma cells [158], human liver cancer cells [159], breast cancer cells [160], gastric cancer xenografts [161], vascular endothelial cells, human bladder cancer P glycoprotein [162], pancreatic islet cells of pancreatic cancerous tissues [163], glioma cell lines [164], etc. These peptide can be expected to be effective bits for tumor targeted therapy and to be of great significance for the early diagnosis of cancer. At the same time, the selected specific biologically active peptides are used as drug carriers and combined with radiochemotherapy drugs for targeted therapy, which provides a new effective treatment for inhibiting tumor cell adhesion, growth, and controlling tumor invasion and metastasis. In our previous work, our laboratory successfully used phage display technology to screen a short peptide TD-1 that can carry protein-based macromolecular drugs to penetrate the skin barrier, and proved that the short peptide can effectively assist insulin through the skin to reduce blood circulation blood glucose. This job provides a good technical platform for the research and lays a solid foundation for this article.

1.6.3.6 Application of Phage Display Technology in Nano Field

Phage display technology uses the interaction between nanomaterials and phage display polypeptides to play an important role in the selection of peptides on the surface of nanomaterials. Early work mainly focused on the synthetic aspects of nanomaterials such as crystallization, nucleation, generation and functional aspects of biomineralization. The earliest Belcher A.M team reported the first use of phage display technology to screen a short peptide GaAs that binds to the surface of semiconductor materials with high affinity, and found that most of the short peptide sequences are polar amino acids [12]. Subsequent research found that for other semiconductor materials and some magnetic materials, such as ZnS [165–167], CdS [165, 167], PbS [165], FePt [167, 168], and CoPt [167], most of the surface display peptides screened were also composed of polar oxyacid residues, indicating that polar amino acids are mainly involved in the interaction between peptides and semiconductor materials, which provides new clues for the synthesis of semiconductor materials. Naik et al. [169] found a short peptide that specifically binds to silver nanoparticels, with the sequence NPSSSLFRYLPSD, named AG4, and found that AG4 can be used as a template to guide the synthesis of silver nanoparticles [170]. A short peptide that specifically binds to nanogold, with the sequence MHGKTQATS-GTIQS, named GBP-1, can be used as a template to control the formation of nanogold [171]. These results provide a way to simulate and study the process of biological mineralization under artificial conditions and provide effective approaches for the development of biocomposites. Another dodecapeptide, MIDAS-2, combined with nano-gold, changes in a single amino acid can regulate the size and morphology of nano-gold, and obtain nano-materials of different sizes and morphologies [172]. Kiyotaka Shiba's group found a specific specificity using phage display technology. The short peptide TBP-1 [173], which binds Ti, Ag and Si, and its biomineralization ability to Si and Ag was studied. Tadafumi Adschiri et al. screened a short peptide ZnO-1 that specifically binds ZnO, and found that cysteine plays an important role in ZnO assembly and biomineralization [174]. However, the study of nanomaterial-specific peptides cannot be stopped only in the level of synthesis and assembly of nanomaterials.

In recent years, with the rapid development of nanomaterials in the field of biomedicine, more and more nanomaterials are used in biological research and disease diagnosis and treatment. Due to the specific selection peptides have high specificity, controllability, and high affinity High biocompatibility, using peptides to promote the application of nanomaterials in biomedicine will become another hot spot in the field of surface display technology. Erica L. Bakota et al. used a peptide that specifically binds carbon nanotubes as a dissolving agent, which could improve the dispersibility of carbon nanotubes, the biocompatibility of the material, and the application of the material in the body [147]. Studies have also shown that peptides can regulate the cell uptake capacity and intracellular toxicity of single-walled carbon nanotubes and nano-amorphous silicon dioxide. Other studies have shown that oligonucleotide-functionalized gold nanoparticles are more easily taken up and internalized by cell membranes. TAT peptides can guide magnetic nanoparticles into cells and promote the application of nanomaterials in core magnetic imaging [175].

This paper will describe another successful example of the application of peptides to promote nanomaterials in biomedicine. We used phage display technology to select a short peptide RE-1 that specifically binds to rare earth nanomaterials, and proved that the short peptide can form a coating on the surface of nanocrystals, which can enhance the suspension ability of nanomaterials and reduce the intercellular interactions, effectively shielding the cellular autophagy activity of the nanomaterial and the toxic effects caused by it. In addition, the bifunctional composite short peptide RE-RGD-1our by RE-1 and arginine glycine aspartate RGD can enhance rare earth up-conversion luminescent nanomaterials by interacting with extracellular integrins. Autophagy and toxic effects in integrin-highly expressed cells show that the goal of shielding autophagy in normal cells and improving autophagy in target cells is expected through the targeting strategy. It can be seen that the biosafety of nanomaterials can be solved through artificial regulation of autophagy effects. At the same time, it is of great significance to systematically study the effects and mechanisms of nanomaterials in regulating autophagy and its relationship with biosafety, establish corresponding detection systems and evaluation standards, and guide the establishment and regulation of nanomaterial production for the establishment of nanomaterial biosafety evaluation systems and safety applications. On the other hand, the use of specific surface-binding peptides to artificially regulate the cellular autophagy behavior of rare earth nanomaterials can greatly reduce the toxic and side effects of nanomaterials while further improving the killing effect on tumor target cells. This makes it possible to achieve the effect of radiotherapy and chemotherapy of cancer cells by regulating the autophagy level of nanomaterials. This result provides new methods and new ideas for the in vivo diagnosis and treatment of rare earth nanomaterials.

References

1. Nie, Z., Petukhova, A., et al. (2010). Properties and emerging applications of self-assembled structures made from inorganic nanoparticles. *Nature Nanotechnology, 5*(1), 15–25.
2. Fahlman, B. D. (2007). *Materials Chemistry, 1*, 282–283 (Springer, Mount Pleasant).
3. Davis, M. E., et al. (2008). Nanoparticle therapeutics: An emerging treatment modality for cancer. *Nature Review Drug Discovery, 7*, 771–782.
4. Robby, A., et al. (2010). Strategies in the design of nanoparticles for therapeutic applications. *Nature Reviews Drug Discovery, 8*, 615–627.
5. Adachi, G., Imanaka, N., & Kang, Z. C. (2004). *Binary Rare Earth Oxides, 1*, 125–126 (Springer, Kluwer, Academic Publishers).
6. Bouzigues, C., Gacoin, T., et al. (2011). Biological applications of rare-earth based nanoparticles. *ACS Nano, 5*(11), 8488–8505.
7. Auzel, F., et al. (2004). Upconversion and anti-stokes processes with f and d ions in solids. *Chem Reviews, 104*(1), 139–173.
8. Bloembergen, N. (1959). Solid state infrared quantum counters. *Physical Review Letters, 2*, 84–85.
9. Auzel, F. (1966). Compteur Quantique Par Transfert D'énergie Entre Deux Ions De Terres Rares Dans Un Tungstate Mixte Et Dans Un Verre. *Comptes Rendus Hebdomadaires Des Seances De L Academie Des Sciences Serie B, 262*, 1016.
10. Yu, M. X., Li, F. Y., Chen, Z. G., Hu, H., Zhan, C., Yang, H., & Huang, C. H. (2009). Laser scanning up-conversion luminescence microscopy for imaging cells labeled with rare-earth nanophosphors. *Analytical Chemistry, 81*(3), 930–935.
11. Xiong, L. Q., Chen, Z. G., Tian, Q. W., Cao, T. Y., Xu, C. J., & Huang, C. H. (2009). High contrast upconversion luminescence targeted imaging in vivo using peptide-labeled nanophosphors. *Analytical Chemistry, 81*(21), 8687–8694.
12. Hu, H., Yu, M. X., Li, F. Y., Chen, Z. G., Gao, X., Xiong, L. Q., & Huang, C. H. (2008). Facile epoxidation strategy for producing amphiphilic up-converting rare-earth nanophosphors as biological labels. *Chemistry of Materials, 20*(22), 7003–7009.
13. Michalet, X., et al. (2005). Quantum dots for live cells, in vivo imaging, and diagnostics. *Science, 307*(5709), 538–544.
14. McCarthy, J. R., et al. (2008). Multifunctional magnetic nanoparticles for targeted imaging and therapy. *Advanced Drug Delivery Reviews, 60*(11), 1241–1251.
15. Gao, J. H., et al. (2009). Multifunctional magnetic nanoparticles: Design, synthesis, and biomedical applications. *Accounts of Chemical Research, 42*(8), 1097–1107.
16. Sharma, C. S., Sarkar, S., et al. (2007). Single-walled carbon nanotubes induces oxidative stress in rat lung epithelial cells. *J Nanoscience and Nanotechnology, 7*(7), 2466–2472.

17. Toma, F. M., Sartorel, A., et al. (2010). Efficient water oxidation at carbon nanotube-polyoxometalate electrocatalytic interfaces. *Nature Chemistry, 2*(10), 826–831.
18. Yamakoshi, Y., Umezawa, N., et al. (2003). Active oxygen species generated from photoexcited fullerene (C60) as potential medicines: O_2^{*} versus $^1O_2$. *Journal of the Amerian Chemical Society, 125*(42), 12803–12809.
19. Corona-Morales, A. A., Castell, A., et al. (2003). Fullerene C60 and ascorbic acid protect cultured chromaffin cells against levodopa toxicity. *Journal of Neuroscience Research, 71*(1), 121–126.
20. Nakanishi, I., Fukuzumi, S., et al. (2002). DNA cleavage via superoxide anion formed in photoinduced electron transfer from NADH to gamma-cyclodextrin-bicapped C-60 in an oxygen-saturated aqueous solution. *Journal of Physical Chemistry, 106*(9), 2372–2380.
21. Adlakha-Hutcheon, G., Bally, M. B., et al. (1999). Controlled destabilization of a liposomal drug delivery system enhances mitoxantrone antitumor activity. *Nature Biotechnology, 17*(8), 775–779.
22. Wing, S. T., et al. (2011). Self-cleaning fibers via nanotechnology: A virtual reality. *Journal of Materials Chemistry, 21*(22), 7858–7869.
23. Deng, Z. J., et al. (2011). Nanoparticle-induced unfolding of fibrinogen promotes Mac-1 receptor activation and inflammation. *Nature Nanotechnology, 6*(1), 39–44.
24. Mikhaylov, G., Mikac, U., et al. (2011). Ferri-liposomes as an MRI-visible drug-delivery system for targeting tumours and their microenvironment. *Nature Nanotechnology, 6*, 594–602.
25. Patra, C. R., Bhattacharya, R., et al. (2008). Pro-angiogenic properties of europium(III) hydroxide nanorods. *Advanced Materials, 20*(4), 753.
26. Hauck, T. S., Anderson, R. E., et al. (2010). In vivo quantum-dot toxicity assessment. *Small (Weinheim an der Bergstrasse, Germany), 6*(1), 138–144.
27. Wang, F., Banerjee, D., et al. (2010). Upconversion nanoparticles in biological labeling, imaging, and therapy. *The Analyst, 135*(8), 1839–1854.
28. Wang, F., Han, Y., et al. (2010). Simultaneous phase and size control of upconversion nanocrystals through lanthanide doping. *Nature, 463*(7284), 1061–1065.
29. Idris, N. M., Li, Z. Q., Ye, L., et al. (2009). Tracking transplanted cells in live animal using upconversion fluorescent nanoparticles. *Biomaterial, 30*(28), 5104–5113.
30. Jiang, S., & Zhang, Y. (2010). Upconversion nanoparticles-based FRET system for study of SiRNA in live cells. *Langmuir, 26*(9), 6689–6694.
31. Xiong, L. Q., Chen, Z. G., et al. (2009). Synthesis, characterization, and in vivo targeted imaging of amine-functionalized rare-earth up-converting nanophosphors. *Biomaterials, 30*(29), 5592–5600.
32. Ivanova, S., Pelle, F., et al. (2008). Upconversion luminescence dynamics of Er-doped fluoride crystals for optical converters. *Journal of Luminescence, 128*(5), 914–917.
33. Suyver, J. F., Grimm, J., et al. (2006). Upconversion spectroscopy and properties of NaYF4 doped with $Er^{(3+)}$, Tm^{3+} and/or Yb^{3+}. *Journal of Luminescence, 117*(1), 1–12.
34. Wang, L. Y., & Li, Y. D. (2006). Green upconversion nanocrystals for DNA detection. *Chemical Communications,* (24), 2557–2559.
35. Lim, S. F., Riehn, R., et al. (2006). In vivo and scanning electron microscopy imaging of upconverting nanophosphors in Caenorhabditis elegans. *Nano Letters, 6*(2), 169–174.
36. Hu, H., Yu, M. X., et al. (2008). Facile epoxidation strategy for producing amphiphilic up-converting rare-earth nanophosphors as biological labels. *Chemistry of Materials, 20*(22), 7003–7009.
37. Chen, Z. G., Chen, H. L., et al. (2008). Versatile synthesis strategy for carboxylic acid-functionalized upconverting nanophosphors as biological labels. *Journal of the Amerian Chemistry Society, 130*(10), 3023–3029.
38. Liu, Z. Y., Yi, G. S., et al. (2008). Monodisperse silica nanoparticles encapsulating upconversion fluorescent and superparamagnetic nanocrystals. *Chemical Communications, 6*, 694–696.

39. Van de Rijke, F., Zijlmans, H., et al. (2001). Up-converting phosphor reporters for nucleic acid microarrays. *Nature Biotechnology, 19*(3), 273–276.
40. Corstjens, P., Zuiderwijk, M., et al. (2001). Use of up-converting phosphor reporters in lateral-flow assays to detect specific nucleic acid sequences: A rapid, sensitive DNA test to identify human papillomavirus type 16 infection. *Clinical Chemistry, 47*(10), 1885–1893.
41. Chatteriee, D. K., Rufalhah, A. J., et al. (2008). Upconversion fluorescence imaging of cells and small animals using lanthanide doped nanocrystals. *Biomaterials, 29*(7), 937–943.
42. Zhang, P., Steelant, W., et al. (2007). Versatile photosensitizers for photodynamic therapy at infrared excitation. *Journal of the American Chemistry Society, 129*(15), 4526.
43. Chatterjee, D. K., & Yong, Z. (2008). Upconverting nanoparticles as nanotransducers for photodynamic therapy in cancer cells. *Nanomedicine, 3*(1), 73–82.
44. Idris, N. M., Gnanasammandhan, M. K., et al. (2012). In vivo photodynamic therapy using upconversion nanoparticles as remote-controlled nanotransducers. *Nature Medicine, 18*(10), 1580–1585.
45. Jiang, S., Zhang, Y., Lim, K. M., Sim, E. K., & Ye, L. (2009). NIR-to-visible upconversion nanoparticles for fluorescent labeling and targeted delivery of siRNA. *Nanotechnology, 20*(15), 155101.
46. Lewinski, N., et al. (2008). Cytotoxicity of nanoparticles. *Small (Weinheim an der Bergstrasse, Germany), 4*(1), 26–49.
47. Nel, A., et al. (2006). Toxic potential of materials at the nanolevel. *Science, 311*(5761), 622–627.
48. Arora, S. (2011). Nanotoxicology and *in vitro* studies: The need of the hour. *Toxicology and Applied Pharmacology, 258*(2), 151–165.
49. Martinet, W., Schrijvers, D. M., et al. (2009). Phagocytosis of bacteria is enhanced in macrophages undergoing nutrient deprivation. *FEBS Journal, 276*(8), 2227–2240.
50. Jaeger, P. A., & Wyss-Coray, T. (2009). All-you-can-eat: Autophagy in neurodegeneration and neuroprotection. *Molecular Neurodegeneration, 4*, 16.
51. Campoy, E., & Colombo, M. I. (2009). Autophagy in intracellular bacterial infection. *Biochimica Et Biophysica Acta-Molecular Cell Research, 1793*(9), 1465–1477.
52. Takeshige, K., Baba, M., et al. (1992). Autophagy in yeast demonstrated with proteinase-deficient mutants and conditions for its induction. *Journal of Cell Biology, 119*(2), 301–311.
53. Matsuura, A., Tsukada, M., et al. (1997). Apglp, a novel protein kinase required for the autophagic process in Saccharomyces cerevisiae. *Gene, 192*(2), 245–250.
54. Mizushima, N., Yoshimori, T., et al. (2011). The role of Atg proteins in autophagosome formation. *Annual Review of Cell and Developmental Biology, 27*, 107–132.
55. Liang, X. H., et al. (1999). Induction of autophagy and inhibition of tumorigenesis by beclin 1. *Nature, 402*, 672–676.
56. Treasures, H. D. N. A. (2004). Area to watch in 2005. *Science, 306*, 2014.
57. Klionsky, D. J. (2007). Autophagy: From phenomenology to molecular understanding in less than a decade. *Nature Reviews Molecular Cell Biology, 8*(11), 931–937.
58. Cuervo, A. M., & Dice, J. F. (1996). A receptor for the selective uptake and degradation of proteins by lysosomes. *Science, 273*(5274), 501–503.
59. Lynch-Day, M. A., Mao, K., et al. (2012). The role of autophagy in Parkinson's disease. *Cold Spring Harbor Perspectives Medicine, 2*(4), a009357.
60. Melendez, A., & Levine, B. (2009). Autophagy in *C. elegans. WormBook, 1*, 1–26.
61. Klionsky, D. T., et al. (2008). Guidelines for the use and interpretation of assays for monitoring autophagy in higher eukaryotes. *Autophagy, 4*(2), 151–175.
62. Klionsky, D. T., et al. (2012). Guidelines for the use and interpretation of assays for monitoring autophagy. *Autophagy, 8*(4), 445–544.
63. Eskelinen, E. L. (2005). Maturation of autophagic vacuoles in mammalian cells. *Autophagy, 1*(1), 1–10.
64. Mizushima, N., Yoshimori, T., et al. (2010). Methods in mammalian autophagy research. *Cell, 140*(3), 313–326.

65. Zheng, Q., Su, H., et al. (2011). Autophagy and p62 in cardiac proteinopathy novelty and significance. *Circulation Research, 109*(3), 296–308.
66. Colosetti, P., Puissant, A., et al. (2009). Autophagy is an important event for megakaryocytic differentiation of the chronic myelogenous leukemia K562 cell line. *Autophagy, 5*(8), 1092–1098.
67. Toepfer, N., Childress, C., et al. (2011). Atorvastatin induces autophagy in prostate cancer PC3 cells through activation of LC3 transcription. *Cancer Biology & Therapy, 12*(8), 691–699.
68. Klionsky, D. J., Cuervo, A. M., et al. (2007). Methods for monitoring autophagy from yeast to human. *Autophagy, 3*(3), 181–206.
69. Pattingre, S., Petiot, A., et al. (2004). Analysis of Gα-interacting protein and activator of G-protein-signaling-3 functions in macroautophagy. *Methods in Enzymology, 390*, 17–31.
70. Bauvy, C., Meijer, A. J., et al. (2009). Assaying of autophagic protein degradation. *Methods in Enzymology, 452*, 47–61.
71. Degenhardt, K., et al. (2006). Autophagy promotes tumor cell survival and restricts necrosis, inflammation, and tumorigenesis. *Cancer Cell, 10*(1), 51–64.
72. Robert, T., et al. (2011). HDACs link the DNA damage response, processing of double-strand breaks and autophagy. *Nature, 471*, 74–79.
73. Liang, X. H., et al. (1999). Induction of autophagy and inhibition of tumorigenesis by beclin 1. *Nature, 402*(6762), 672–676.
74. Mathew, R., et al. (2007). Autophagy suppresses tumor progression by limiting chromosomal instability. *Genes and Development, 21*, 1367–1381.
75. Mathew, R., et al. (2009). Autophagy suppresses tumorigenesis through elimination of p62. *Cell, 137*, 1062–1075.
76. Xie, Z., et al. (2007). Autophagosome formation: Core machinery and adaptations. *Nature Cell Biology, 9*(10), 1102–1109.
77. Youle, R. J., et al. (2011). Mechanisms of mitophagy. *Nature Reviews Molecular Cell Biology, 12*(1), 9–14.
78. Guo, J., et al. (2011). Activated Ras requires autophagy to maintain oxidative metabolism and tumorigenesis. *Genes and Development, 25*(5), 460–470.
79. Yang, S., et al. (2011). Pancreatic cancers require autophagy for tumor growth. *Genes and Development, 25*(7), 17–29.
80. Hu, Y., et al. (2012). Hypoxia-induced autophagy promotes tumor cell survival and adaptation to antiangiogenic treatment in glioblastoma. *Cancer Research, 72*(7), 1773–1783.
81. Amaravadi, R., et al. (2011). Principles and current strategies for targeting autophagy for cancer treatment. *Clinical Cancer Research, 17*(4), 654–666.
82. Yang, Z., et al. (2011). The role of autophagy in cancer: Therapeutic implications. *Molecular Cancer Therapeutics, 10*(9), 1533–1541.
83. Matsuda, F., et al. (2009). Autophagy induced by 2-deoxy-D-glucose suppresses intracellular multiplication of Legionella pneumophila in A/J mouse macrophages. *Autophagy, 5*(4), 484–493.
84. Mizushima, N., et al. (2008). Autophagy fights disease through cellular self-digestion. *Nature, 451*(7182), 1069–1075.
85. Lee, J. H., et al. (2010). Lysosomal proteolysis and autophagy require presenilin 1 and are disrupted by Alzheimer-related PS1 mutations. *Cell, 141*(7), 1146–1191.
86. Wong, E. S., et al. (2008). Autophagy-mediated clearance of aggresomes is not a universal phenomenon. *Human Molecular Genetics, 7*(16), 2570–2582.
87. Vives-Bauza, C., de Vries, R. L., et al. (2010). PINK1/Parkin direct mitochondria to autophagy. *Autophagy, 6*(2), 315–316.
88. Hara, T., et al. (2006). Suppression of basal autophagy in neural cells causes neurodegenerative disease in mice. *Nature, 441*(7095), 885–889.
89. Komatsu, M., et al. (2006). Loss of autophagy in the central nervous system causes neurodegeneration in mice. *Nature, 441*(7095), 880–884.
90. Tsukamoto, S., et al. (2008). Autophagy is essential for preimplantation development of mouse embryos. *Science, 321*(5885), 117–120.

91. Al Rawi, S., et al. (2011). Postfertilization autophagy of sperm organelles prevents paternal mitochondrial DNA transmission. *Science, 334*(6059), 1144–1147.
92. Qu, X., et al. (2007). Autophagy gene-dependent clearance of apoptotic cells during embryonic development. *Cell, 128*(5), 931–946.
93. Kuma, A., et al. (2004). The role of autophagy during the early neonatal starvation period. *Nature, 432*(7020), 1032–1036.
94. Paludan, C., et al. (2005). Endogenous MHC class II processing of a viral nuclear antigen after autophagy. *Science, 307*(5709), 593–596.
95. Nedjic, J., et al. (2008). Autophagy in thymic epithelium shapes the T-cell repertoire and is essential for tolerance. *Nature, 455*(7211), 396–400.
96. Nakagawa, I., et al. (2004). Autophagy defends cells against invading group A Streptococcus. *Science, 306*(5698), 1037–1040.
97. Lee, H. K., et al. (2007). Autophagy-dependent viral recognition by plasmacytoid dendritic cells. *Science, 315*(5817), 1398–1401.
98. Singh, S. B., et al. (2006). Human IRGM induces autophagy to eliminate intracellular mycobacteria. *Science, 313*(5792), 1438–1441.
99. Munz, C. (2010). Antigen processing via autophagy—not only for MHC class II presentation anymore? *Current Opinion Immunology, 22*(1), 89–93.
100. Levine, B., et al. (2008). Autophagy in the pathogenesis of disease. *Cell, 132*(1), 27–42.
101. Ogawa, M., et al. (2005). Escape of intracellular Shigella from autophagy. *Science, 307*(5710), 727–731.
102. Wileman, T. (2006). Aggresomes and autophagy generate sites for virus replication. *Science, 312*(5775), 875–878.
103. Virgin, H. W., et al. (2009). Autophagy genes in immunity. *Nature Immunology, 10*(5), 461–470.
104. Taylor, M. P., & Kirkegaard, K. (2007). Modification of cellular autophagy protein LC3 by poliovirus. *Journal of Virology, 81*(22), 12543–12553.
105. Brabec-Zaruba, M., et al. (2007). Induction of autophagy does not affect human rhinovirus type 2 production. *Journal of Virology, 81*(19), 10815–10817.
106. Blanchet, F. P., Moris, A., et al. (2010). Human immunodeficiency virus-1 inhibition of immunoamphisomes in dendritic cells impairs early innate and adaptive immune responses. *Immunity, 32*(5), 654–669.
107. Nicklin, P., et al. (2009). Bidirectional transport of amino acids regulates mTOR and autophagy. *Cell, 136*(3), 521–534.
108. Singh, R., et al. (2009). Autophagy regulates lipid metabolism. *Nature, 458*(7242), 1131–1135.
109. He, C., et al. (2012). Exercise-induced BCL2-regulated autophagy is required for muscle glucose homeostasis. *Nature, 481*(7382), 511–515.
110. Rubinsztein, D. C., et al. (2011). Autophagy and aging. *Cell, 146*(5), 682–695.
111. Green, D. R., et al. (2011). Mitochondria and the autophagy-inflammation-cell death axis in organismal aging. *Science, 333*(6046), 1109–1112.
112. Melendez, A., et al. (2003). Autophagy genes are essential for dauer development and life-span extension in *C. elegans. Science, 301*(5638), 1387–1391.
113. Johnson,T. E. (2008). Caenorhabditis elegans 2007: The premier model for the study of aging. *Experimental Gerontology, 43*(1), 1–4.
114. Chen, Y., et al. (2005). Nano neodymium oxide induces massive vacuolization and autophagic cell death in non-small cell lung cancer NCI-H460 cells. *Biochemical Biophysical Research Commun, 337*(1), 52–60.
115. Seleverstov, O., et al. (2006). Quantum dots for human mesenchymal stem cells labeling. A size-dependent autophagy activation. *Nano Letter, 6*(12), 2826–2832.
116. Stern, S. T., et al. (2008). Induction of autophagy in porcine kidney cells by quantum dots: A common cellular response to nanomaterials? *Toxicological Sciences, 106*(1), 140–152.
117. Zhang, Q., et al. (2009). Autophagy-mediated chemosensitization in cancer cells by fullerene C60 nanocrystal. *Autophagy, 5*(8), 1107–1117.

118. Yu, L., et al. (2009). Rare earth oxide nanocrystals induce autophagy in HeLa cells. *Small (Weinheim an der Bergstrasse, Germany), 5*(24), 2784–2787.
119. Man, N., et al. (2010). Rare earth oxide nanocrystals as a new class of autophagy inducers. *Autophagy, 6*(2), 310–311.
120. Liu, H. L., et al. (2011). A functionalized single-walled carbon nanotube-induced autophagic cell death in human lung cells through Akt-TSC2-mTOR signaling. *Cell Death and Disease, 2*, e159.
121. Wei, P., et al. (2010). C60 (Nd) nanoparticles enhance chemotherapeutic susceptibility of cancer cells by modulation of autophagy. *Nanotechnology, 21*(49), 495101–495112.
122. Lee, C., et al. (2011). C60 fullerene-pentoxifylline dyad nanoparticles enhance autophagy to avoid cytotoxic effects caused by the β-amyloid peptide. *Nanomedicine, 7*(1), 107–114.
123. Zhang, Y., et al. (2010). Nano rare-earth oxides induced size-dependent vacuolization: An independent pathway from autophagy. *International Journal of Nanomedicine, 5*, 601–609.
124. Khan, M. I., et al. (2012). Induction of ROS, mitochondrial damage and autophagy in lung epithelial cancer cells by iron oxide nanoparticles. *Biomaterials, 33*(5), 1477–1488.
125. Ma, X., et al. (2011). Gold nanoparticles induce autophagosome accumulation through size-dependent nanoparticle uptake and lysosome impairment. *ACS Nano, 5*(11), 8629–8639.
126. Wu, Y. N., et al. (2011). The selective growth inhibition of oral cancer by iron core-gold shell nanoparticles through mitochondria-mediated autophagy. *Biomaterials, 32*(20), 4565–4573.
127. Li, C., et al. (2009). PAMAM nanoparticles promote acute lung injury by inducing autophagic cell death through the Akt-TSC2-mTOR signaling pathway. *Journal of Molecular Cell Biology, 1*(1), 37–45.
128. Man, N., et al. (2010). Induction of genuine autophagy by cationic lipids in mammalian cells. *Autophagy, 6*(4), 449–454.
129. Eidi, H., et al. (2012). Drug delivery by polymeric nanoparticles induced autophagy in macrophages. *International Journal of Pharmaceutics, 422*(1), 495–503.
130. Kroemer, G., et al. (2008). Autophagic cell death: The story of a misnomer. *Nature Reviews Molecular Cell Biology, 9*(12), 1004–1010.
131. Chiara Maiuri, M. (2007). Self-eating and self-killing: Crosstalk between autophagy and apoptosis. *Nature Reviews Molecular Cell Biology, 8*(9), 741–752.
132. Yang, Lu., et al. (2012). MnO nanocrystals: A platform for integration of MRI and genuine autophagy induction for chemotherapy. *Advanced Functional Materials, 10*, 1002.
133. Li, J. J., Hartono, D., et al. (2010). Autophagy and oxidative stress associated with gold nanoparticles. *Biomaterials, 31*(23), 5996–6003.
134. Li, H. Y., Li, Y. H., Jiao, J., & Hu, H. M. (2011). Alpha-alumina nanoparticles induce efficient autophagy-dependent cross-presentation and potent antitumor response. *Nature Nanotechnology, 6*(10), 645–650.
135. Chiu, H. (2011). Synergistic effects of arsenic trioxide and radiation in osteosarcoma cells through the induction of both autophagy and apoptosis. *Radiation Research, 175*(5), 547–560.
136. Johnson-Lyles, D., et al. (2010). Fullerenol cytotoxicity in kidney cells is associated with cytoskeleton disruption, autophagic vacuole accumulation, and mitochondrial dysfunction. *Toxicology and Applied Pharmacology, 248*(3), 249–258.
137. Ueng, T., et al. (1997). Suppression of microsomal cytochrome P450-dependent monooxygenases and mitochondrial oxidative phosphorylation by fullerenol, a polyhydroxylated fullerene C60. *Toxicology Letters, 93*(1), 29–37.
138. Chen, H., et al. (1998). Acute and subacute toxicity study of water-soluble polyalkylsulfonated C60 in rats. *Toxicologic Pathology, 26*(1), 143–151.
139. Monick, M., et al. (2010). Identification of an autophagy defect in smokers' alveolar macrophages. *Journal of Immunology, 185*(9), 5425–5435.
140. Chen, M. Y., et al. (2005). Surface properties, more than size, limiting convective distribution of virus-sized particles and viruses in the central nervous system. *Journal of Neurosurgery, 103*(2), 311–319.
141. Zhang, X. L., Niu, H. Y., et al. (2010). Chitosan-coated octadecyl-functionalized magnetite nanoparticles: preparation and application in extraction of trace pollutants from environmental water samples. *Analytical Chemistry, 82*(6), 2363–2371.

142. Chen, Y. J., Tao, J. A., et al. (2010). Synthesis, self-assembly, and characterization of PEG-coated iron oxide nanoparticles as potential MRI contrast agent. *Drug Development and Industrial Pharmacy, 36*(10), 1235–1244.
143. Alam, M. M., et al. (2010). Synthesis and characterization of TTAB coated Silver (Ag) nanoparticles. *Advanced Materials Research, 264*, 530–534.
144. Whaley, S. R., English, D. S., Hu, E. L., Barbara, P. F., & Belcher, A. M. (2000). Selection of peptides with semiconductor binding specificity for directed nanocrystal assembly. *Nature, 405*(6787), 665–668.
145. Brown, S., Sarikaya, M., et al. (2000). A genetic analysis of crystal growth. *Journal of Molecular Biology, 299*(3), 725–735.
146. Giljohann, D. A., Seferos, D. S., et al. (2007). Oligonucleotide loading determines cellular uptake of DNA-modified gold nanoparticles. *Nano Letters, 7*(12), 3818–3821.
147. Bakota, E. L., Aulisa, L., Tsyboulski, D. A., Weisman, R. B., & Hartgerink, J. D. (2009). Multidomain peptides as single-walled carbon nanotube surfactants in cell culture. *Biomacromolecules, 10*(8), 2201–2206.
148. Kim, Y. H., Jeon, J., et al. (2011). Tumor targeting and imaging using cyclic RGD-PEGylated gold nanoparticle probes with directly conjugated iodine-125. *Small (Weinheim an der Bergstrasse, Germany), 7*(14), 2052–2060.
149. Dutta, D., et al. (2007). Adsorbed proteins influence the biological activity and molecular targeting of nanomaterials. *Toxicological Sciences, 100*(1), 303–315.
150. Smith, G. P. (1985). Filamentous fusion phage: Novel expression vectors that display cloned antigens on the virion surface. *Science, 228*, 1315–1317.
151. Barbas, C. F., III, Burton, D. R., Scott, J. K., & Silverman, G. J. (2001). *Phage display: A laboratory manual* (Cold Spring Harbour Laboratory Press, New York).
152. Lademann, J., et al. (1999). Penetration of titanium dioxide microparticles in a sunscreen formulation into the horny layer and the follicular orifice. *Skin Pharmacology and Applied Skin Physiology, 12*, 247–256.
153. Lowman, H. B., et al. (1991). Selecting high-affinity binding proteins by monovalent phage display. *Biochemistry, 30*(45), 10832–10838.
154. Pasqualini, R., & Ruoslahti, E. (1996). Organ targeting in vivo using phage display peptide libraries. *Nature, 380*, 364–366.
155. Rajotte, D., et al. (1998). Molecular heterogeneity of the vascular endothelium revealed by in vivo phage display. *The Journal of Clinical Investigation, 102*(2), 430–437.
156. Bai, F., Liang, J., et al. (2007). Inhibitory effects of a specific phage-displayed peptide on high peritoneal metastasis of gastric cancer. *Journal of Molecular Medcine, 85*, 169–180.
157. Wang, S. X., et al. (2003). Selection of targets locating on the surface of epithelial ovarian cancer cells by using phage peptide library. *Zhonghua Fu Chan Ke Za Zhi, 38*(7), 412–414.
158. Hsiung, P. L., et al. (2008). Detection of colonic dysplasia in vivo using a targeted heptapeptide and confocal microendoscopy. *Nature Medicine, 14*(4), 454–458.
159. Du, Y. Z., Cai, L. L., et al. (2012). Tumor cells-specific targeting delivery achieved by A54 peptide functionalized polymeric micelles. *Biomaterials, 33*(34), 8858–8867.
160. Shukla, G. S., et al. (2005). Selection of tumor-targeting agents on freshly excised human breast tumors using a phage display library. *Oncology Reports, 13*(4), 757–764.
161. Wang, Y., et al. (2003). In vivo screening and characterization of peptides specific binding to vasculature of gastric cancer. *Chinese Journal of Cell Molmmunology, 19*(5), 469–472.
162. Wang, X. W., et al. (2003). Biospanning of peptide specific to P glycopmtein. *Chinese Journal of Urology, 24*(12), 824–826.
163. Yao, V. J., et al. (2005). Targeting pancreatic islets with phage display assisted by laser pressure catapult microdissection. *The American Journal of Pathology, 166*(2), 625–636.
164. Wang, B., et al. (2004). Selection of targeted glioblastoma tumor cell binding and internalization peptides through phage display vector. *Chinese Journal of Pathophysiology, 20*(5), 752–756.
165. Flynn, C. E., Mao, C. B., et al. (2003). Synthesis and organization of nanoscale II-VI semiconductor materials using evolved peptide specificity and viral capsid assembly. *Journal of Materials Chemistry, 13*(10), 2414–2421.

166. Lee, S. W., Mao, C. B., et al. (2002). Ordering of quantum dots using genetically engineered viruses. *Science, 296*(5569), 892–895.
167. Mao, C. B., Solis, D. J., et al. (2004). Virus-based toolkit for the directed synthesis of magnetic and semiconducting nanowires. *Science, 303*(5655), 213–217.
168. Reiss, B. D., Mao, C. B., et al. (2004). Biological routes to metal alloy ferromagnetic nanostructures. *Nano Letters, 4*(6), 1127–1132.
169. Naik, R. R., Brott, L. L., et al. (2003). Bio-inspired approaches and biologically derived materials for coatings. *Progress in Organic Coatings, 47*(3), 249–255.
170. Lee, E., Lim, Y., et al. (2008). Solution structure of peptide AG4 used to form silver nanoparticles. *Biochemical adn Biophysical Research Communications, 376*(3), 595–598.
171. Kulp, J. L., Sarikaya, M., et al. (2004). Molecular characterization of a prokaryotic polypeptide sequence that catalyzes Au crystal formation. *Journal of Materials Chemistry, 14*(14), 2325–2332.
172. Kim, J., et al. (2010). Peptide-mediated shape- and size-tunable synthesis of gold nanostructures. *Acta Biomater ialia, 6*(7), 2681–2689.
173. Sano, K. I., Sasaki, H., et al. (2005). Specificity and biomineralization activities of Ti-binding peptide-1 (TBP-1). *Langmuir, 21*(7), 3090–3095.
174. Okochi, M., Sugita, T., et al. (2010). Peptide array-based characterization and design of ZnO-high affinity peptides. *Biotechnology and Bioengineering, 106*(6), 845–851.
175. Lewin, M., et al. (2000). Tat peptide-derivatized magnetic nanoparticles allow in vivo tracking and recovery of progenitor cells. *Nature Biotechnology, 18*(4), 410–414.

Chapter 2
Successfully Obtained Short Peptides RE-1 Using Phage Display Technology

2.1 Introduction

Phage display technology uses the interaction between nanomaterials and phage display polypeptides, playing an important role in the selection of peptides on the surface of nanomaterials. At present, there have been many successful cases of using phage display technology to obtain material-specific binding peptides and promote the application of materials. The peptides specifically bound to nanomaterials screened by phage display technology can be combined with high affinity to specific peptides by simple mixing. This binding is different from non-specific adhesion with lower affinity and different from chemical covalent binding, which can successfully achieve high-affinity binding only under mild conditions.

Therefore, it is possible to obtain short peptides by phage display technology that can specifically bind rare earth nanomaterials, change their surface properties and physical and chemical properties, shift their biological effects, and promote their biomedical applications. In this chapter, we will describe the first use of phage display technology to obtain a short peptide RE-1 that specifically binds to rare earth nanomaterials, and further identify its binding ability.

2.2 Experimental Materials

2.2.1 Reagents

- Phage library kit (PH.D.-C7C Phage Display Peptide Library Kit, NEW ENGLAND Biolands, Inc.) purchased from New England Biotechnology. The kit includes the phage library, primers and host bacteria necessary for the experiment, as follows:

Y. Zhang, *Tuning Autophagy-Inducing Activity and Toxicity for Lanthanide Nanocrystals*, Springer Theses, https://doi.org/10.1007/978-981-16-8166-0_2

1. A random heptapeptide phage display library (PH.D.-C7C), stored in a solution containing 50% glycerol, contains about 2.8×10^9 transformants.
2. -28 gIII primers for sequence
3. -^{HO} GTA TGG GAT TTT GCT AAA CAA C-3', 100 pm, 1 pmol/uL
4. -96 gIII primers for sequence
5. -^{HO} CCC TCA TAG TTA GCG TAA CG-3', 100 pm, 1 pmol/uL
6. E.coli ER2738 library.

- LB liquid culture medium: per liter: 10 g peptone, 5 g yeast extract, 5 g NaCl, to 1L, sterilized at 121 °C for 20 min and stored at room temperature.
- LB solid culture medium: per liter: 15 g agarose, 10 g peptone, 5 g yeast extract, 5 g NaCl, to 1L, sterilized at 121 °C for 20 min and stored at room temperature.
- Agarose Top medium: per liter: 7 g agarose, 10 g peptone, 5 g yeast extract, 5 g NaCl, 1 g $MgCl_2 \cdot 6H_2O$, divide into 50 mL per bottle, sterilize at 121 °C for 20 min and store at room temperature.
- IPTG/X gel Solution: Dissolve 1.25 g IPTG and 1 g X gel in 25 dimethylformamide. After mixing, distribute into 200 uL per tube, and store them in the—20 °C refrigerator protected from light.
- LB/IPTG/X gel Plate: LB solid medium, sterilized at high temperature, cooled to below −70 °C, add 1 mL IPTG/X gel solution, and spread the plate. Store the plate at 4 °C in a dark place.
- Tetracycline: 100 g powder is weighed and dissolved in 5 mL ethanol, prepared as 20 mg/mL solution, divided into 200 uL per tube, store them in the—20 °C refrigerator protected from light.
- LB/Tetracycline plate: When the sterilized LB solid medium is cooled to 45–50 °C, add the tetracycline solution according to the proportion 1:1000 and pour the plate. After the plate has cooled, store it at 4 °C in an inverted position. Note: The temperature of the culture medium should not be too high when inverting the plate. The high temperature will invalidate the tetracycline. In addition, the plate must be thoroughly cooled before being placed in the 4°, otherwise a large amount of condensate will be collected.
- LB/Tetracycline liquid culture medium: Before use, add the tetracycline solution to the liquid culture medium in accordance with the ratio 1:1000 to use.
- TBS: Prepare 250 mL 50 mL tris–HCl, 150 mM NaCl into 500 mL wide mouth bottles and store at room temperature after sterilization.
- 2% TBST Washing liquid: Adding TBS to 2% tween-20 and mixing evenly, and store at room temperature after sterilization.
- PEG/NaCl Preparation: Prepare 250 mL 20% PEG-8000, 2.5 mM NaCl into 50 ml wide mouth bottles and store at room temperature after sterilization.
- TE Buffer: Prepare 25 mL 10 mM tris–HCl, 1 mM EDTA into 50 ml centrifuge tube, sterilize, and store at room temperature.
- Sodium acetate/ ethanol solution: Prepare 0.3 M sodium acetate: ethanol = 1:20; the final volume is 100 mL, put into 250 mL bottle, and store at room temperature.

- Glycine hydrochloric acid eluent: Make up 0.2 M Glycine–HCl, 1 mg/mL BSA to 50 mL. Since BSA are sensitive to high temperature sterilization, in order to prevent bacterial contamination, the eluent should be prepared as needed.
- Nano-Nd_2O_3 suspension: The nano-powder was purchased from Guangzhou Huizhou Ruier Chemical. The powder was weighed 10 mg and formulated into 1 mg/ml, store into 50 mL centrifuge tube, placed in an ultrasonic cell disruptor, and treated at room temperature.
- YCl3, YbCl3, NH4F were purchased from the company; Methanol, oleic acid, stearyl, ethanol, cyclohexane, etc. were purchased from Reagent library of University of Science and Technology of China.
- 2% Phosphoric acid: weigh 1 g phosphatidic acid powder, formulated into 2%. using potassium hydroxide to adjust PH to 6.7 and store at 4 °C.

2.2.2 *Experimental Equipment and Consumables*

Triangular flasks, beakers, inoculation rings, flat glass dishes, alcohol lamps, copper mesh sample boxes were purchased from the Chemical Reagent Library of the University of Science and Technology of China.

Parafilm (BEMIS, United States).

2.3 Instruments and Equipment

Scourge Mixer (China).

37 °C Constant temperature shaker (Shanghai Zhicheng, China)
4 °C Refrigerator (Siemens, Germany)
−20 °C Freezer (Aucma, China)
−80 °C Ultra-low temperature refrigerator (Zhongke Meiling, China)
Constant temperature water bath (Jintan Ronghua, China)
Electronic balance (Switzerland)
Biochemical Incubator (Shanghai Yuejin Medical Equipment Factory, China)
Milli-Q Ultrapure water purification system (USA)
Microcentrifuge (Germany)
Ultra-low temperature centrifuge (Germany)
Ultra Clean Workbench (Shanghai Chengshun, China)
HPLC (United States)
Sterilizer (Japan)
PCR Instrument (Germany)
Electrophoresis instrument (China)
Gel imaging system (China)
Ultrasonic Crusher (Nanjing Xianou, China)

2D-MS Mass Spectrometer (USA)
UV Spectrophotometer (US)
Transmission electron microscope (Japan)

2.4 Experimental Method

2.4.1 *Screening REOB-1 by Phage Display Technology*

2.4.1.1 Activate Host Strain: *E. coli* ER2738

Burn the sterilized inoculation ring on the alcohol lamp and cool it on the plate. Dip a small amount of the host bacteria bacterial solution in the kit, inoculate it on the tetracycline plate, and invert and incubate at 37 °C. When it is backward, the plate is sealed with a parafilm and stored in a 4° refrigerator for future use.

2.4.1.2 Cultivate Host Strain

A single clone from the freshly activated host bacteria plate was inoculated into a liquid tetracycline medium and shake-cultured in a 37 °C constant temperature shaker. When the bacterial solution reached the logarithmic growth phase during amplification, it was measured by a spectrophotometer. Amplification can be stopped when the OD600 value is 0.5, and the amplified bacterial solution is dispensed into a sterilized 15 centrifuge tube for later use.

2.4.1.3 Amplified M13phage

Inoculate a single 50-ml Erlenmeyer flask containing 10 ml of liquid culture medium, add tetracycline (ratio 1: 1000), and incubate at 37 °C with a shaker overnight. When the bacterial solution is expanded to a saturated concentration, adding 1–2 ml to 500 ml volume erlenmeyer flask containing 100 ml of LB liquid culture medium, inserting into the sputum body to be amplified and cultured with shaking in a 37° constant temperature shaker. After 5 h of incubation, transfer the supernatant to a clean centrifuge tube by centrifugation, add 1/6 volume of PEG/NaCl solution, and mix well and put it in a 4 °C refrigerator overnight. The next day, the above-mentioned precipitated solution was centrifuged at 14,000 rpm at 4 °C and the supernatant was discarded. The pellet was resuspended in 1 ml of sterilized TBS. After the pellet was completely dissolved, centrifuge again to remove impurities and transfer the supernatant to a centrifuge tube. Add 1/6 volume of solution to precipitate the phage again, and ice bath for 60 min. Centrifuge at 4° for 10 min, discard the supernatant, and the pellet is phage particles. Resuspend the pellet in sterilized TBS until dissolved. Store in a refrigerator at 4° until the titer is determined.

2.4.1.4 Measuring Phage Titer

Place the LB/IPTG/X gel plate at 37° in advance for standby at least 1 h, and then it can be removed before use. After the Top medium is thawed by microwave heating, aliquot it into centrifuge tubes, add about 3 ml to each tube, and place in a water bath for later use. Take 1 ul of the amplified phage solution and dilute it to four concentration gradients of 10^{-7}, 10^{-8}, 10^{-9}, 10^{-10}. Take 10 ul of each and add the host bacterial solution in logarithmic phase, mix well and incubate at room temperature for 1–5 min. Add the mixture to the top liquid medium in the above 45° water bath, mix thoroughly and quickly pour it on the pre-heated LB / IPTG / X gel plate. Tilt and rotate the plate to evenly spread the Top medium on the plate, lay it flat on a clean bench, cool it, and invert it in a 37° bacteria incubator overnight. The next day, the phage titer was calculated based on the number of blue plaques sprayed on each plate and the dilution factor.

2.4.1.5 Three Rounds of Screening for Phages that Specifically Bind to Nano-Nd_2O_3

Take a total of 10^{11} phage libraries and mix them with a nano-Nd_2O_3 suspension at a concentration of 1 mg/ml and mix them with shaking in a constant temperature shaker at 37 °C. After centrifugation at high speed for 10 min, discard the supernatant and wash the pellet 10 times. The unbound phage was removed and the bound phage was recovered with 0.2 M Glycine–HCl eluent. The titer and amplification of the phage recovered by elution were measured, and the second round of screening was performed. Take the first round of the amplified product and mix it with the nano-Nd_2O_3 suspension, and shake and incubate it in a 37° constant temperature shaker for 2 h. Repeat the above steps to remove unbound phage, and recover the second enriched bacteriophages. In order to obtain highly specific bound phages, we conducted a third round of screening. The method is exactly the same as the first and second rounds of screening. The three rounds of enriched phages were collected and the titers were measured.

2.4.1.6 Pick a Monoclonal

The recovered product from the third round of screening was diluted to a certain proportion and spread on a LB/IPTG/X gel plate and placed in a 37° bacteria incubator for inversion. The next day, the medium (containing tetracycline) was aliquoted into a single monoclonal amplification tube. The culture plate with blue plaque was taken out of the incubator, and 15 monoclonal chewing bacteria were randomly picked from multiple plates. The body was filled into monoclonal amplification tubes, and 30ul of bacterial solution was added for amplification. Then collect the phage particles in each amplification tube, mark the titer after labeling, and reserve.

2.4.1.7 Extraction of Phage DNA

Take the bacterial solution of the amplified phage and centrifuge at 4° to transfer the supernatant to a new centrifuge tube. Add 1/6 volume PEG/NaCl solution to the centrifuge tube, invert the centrifuge tube upside down and mix by gentle shaking, and let stand at room temperature for 15–30 min. Centrifuge at 4° for 10 min, discard the supernatant, and recover the bacterial particles in the pellet. Resuspend the phage pellet and shake to completely dissolve the pellet. Add balance buffer, shake and mix for 30 s, and let it stand at room temperature before shaking to make it mix well. Centrifuge the two phases at room temperature and transfer as much of the aqueous phase as possible to a new centrifuge tube. Add 2 volumes of ethanol and mix for 15–30 min at room temperature. Centrifuge to recover the single-stranded phage DNA pellet. Gently aspirate the supernatant and centrifuge briefly to completely remove the residual supernatant. Add cold ethanol to the pellet to carefully wash the pellet and centrifuge at 4°. Immediately aspirate and discard the supernatant. Open the tube cap of the centrifuge tube, and turn the cap upside down to accelerate air circulation to completely remove the residual ethanol. TE is added to the centrifuge tube to dissolve the DNA precipitate. At this time, the centrifuge tube can be incubated in a 37° incubator to accelerate the dissolution. The dissolved DNA solution is stored in a -20° refrigerator for future use.

2.4.1.8 Phage DNA Sequence Determination

The 15 monoclonal phage DNA solutions and random monoclonal phage DNA solutions extracted above were sent to Invitrogen for sequencing.

2.4.2 Design and Synthesis of Monoclonal Phage-Specific Primers

According to the sequencing results, the monoclonal phage-specific primers were designed according to the sequence of the peptide displayed on the surface of the phage, and sent to Shanghai Shengong Biological Company to synthesize primers.

2.4.3 PCR Identification of Specific Phage Binding Capacity

Mix REOB-1 phage with a titer of 10^5 and AP-1 phage (random phage) with a titer of 10^8, mix it with a nano-Nd_2O_3 suspension with a concentration of 1 mg/ml, and incubate in a 37° C shaker with high speed. After centrifugation for 10 min, the supernatant was discarded, and the pellet was washed 10 times with TBST. The unbound phage was washed away, and then the phage bound to the nano-Nd_2O_3 was eluted with 0.2 M Glycine HCl eluent, and diluted to a certain extent and spread on LB paltes. Fifty blue plaques were randomly picked on multiple plates, and each plaque was divided into two parts for simultaneous double PCR reaction. One half was reacted with primers specific to REOB-1 phage, and the other half was reacted with primers specific to AP-1 phage. The additional primers required for the PCR reaction consist of sequences on the vector and are shared by both reactions. The agarose gel electrophoresis was performed on the PCR products, and the electrophoresis results were photographed with a gel imaging system.

2.4.4 Chemical Synthesis of RE-1 and Its Analogs

The peptides used in this article were used by Shanghai Gill Biochemical Company.

The peptide automatic synthesizer was obtained by standard solid-phase FMOC synthesis method, and its purity was more than 99% by HPLC analysis. All synthesized peptides were identified by mass spectrometry and confirmed to be correct before use.

2.4.5 UCN Synthesis Method

The rare earth up-conversion luminescent nanomaterials used in this paper are spherical $NaYF_4$: 18% Yb, 2% Er nanoparticles, and the synthesis method is as follows. YCl_3, $YbCl_3$, $ErCl_3$ are mixed with 6 mL oleic acid and 17 mL eighteen dilute in a flask, and heated to 160 degrees Celsius. Then cooled to room temperature. NaOH and NH_4F were dissolved in 10 mL of methanol and added dropwise to the above flask, and stirred for 30 min to ensure that the chlorides fully reacted. The above solution was slowly heated to evaporate the methanol, degassed at 100 °C for 10 min, and then heated to 300° in argon for 1 h. The solution was allowed to cool naturally, at which point the nanocrystals precipitated. It was then precipitated with ethanol and washed three times, and then treated with 1 M HCl for 5 h at room temperature to remove the oleic acid on the surface of the nanoparticles, and washed 10 times with water and cyclohexane. The ultraviolet absorption spectrophotometer was used to detect the characteristic absorption peak of oleic acid at 230 nm to ensure that oleic acid was completely removed. In the synthesis process, the size of UCN and UCN-S nanoparticles was controlled by adding different amounts of oleic acid. The synthesized UCN was stored in water for experimental use.

2.4.6 TEM Observation of Phage REOB-1 Binding to UCN

Transmission electron microscope was used to observe the binding form of phage REOB-1 to UCN. The phage REOB-1 was mixed with 100 ug UCN and incubated with shaking in a 37° constant temperature shaker. Then, the nanoparticles precipitated after centrifugation were washed with water several times until the unbound phage was removed. Because the phage is a biological sample, the contrast of the sample itself is extremely poor under the electron microscope, and the background of the sample needs to be stained with a staining agent using a negative staining method to highlight the sample. Generally, a substance with a high electron density that does not show any structure itself and does not react with the sample. Such as the sodium salt of phosphotungstic acid, it surrounds the sample, and at the same time, the staining agent is also injected into the observation object. The shape can reflect the internal structure of the sample to a certain extent, so as to show a clear ultrastructure under the electron microscope. Therefore, the binding products and phages need to be stained with phosphotungstic acid for observation. UCN does not require staining for direct observation. The carbon film-coated copper mesh was clamped on the filter paper with special tweezers for electron microscopy, with the reflective surface of the copper mesh facing downwards, and the above-mentioned stained sample and UCN suspension were gently dripped onto the copper mesh with a micropipette to prevent reflection. On one side, after the sample is evenly dispersed on the copper mesh, take a clean glass dish and cover the copper mesh to avoid dust pollution in the air. After the sample was air-dried, the image was observed using a transmission electron microscope, and the results were recorded. The observed copper mesh was placed in a copper mesh sample box for storage.

2.4.7 Identification of Binding Ability of Peptide Analogs to Nd_2O_3

Mix nano-Nd_2O_3 with RE-1, AP-1, RE-2, RE-3, RE-4, RE-5, RE-6, RE-7, RE-8, RE-9, RE-10 separately, and then add REOB-1 phage. After mixing, shake and incubate in a 37° temperature shaker for 2 h. After centrifugation at high speed for 10 min, discard the supernatant, wash the pellet with TBST 10 times, wash away unbound phage, and the phages bound to nano-Nd_2O_3 were eluted and the titer was measured.

2.4.8 Competitive Binding Experiments of RE-1 Short Peptide Phages with Nano-Nd_2O_3

Mix short peptides with different concentration gradients with Nd_2O_3 and REOB-1, vortex and shake, incubate at 37° in a constant temperature shaker, discard the supernatant, wash the pellet 10 times with TBST, and wash away unbound phage. Then, the phage bound to the nanometer was eluted with an eluent, and the titer of the recovered phage was measured in each tube. AP-1 short peptide was used as a control.

2.4.9 Data Analysis

All experiments were repeated at least three times. All data are expressed as mean ± S.E.M., Students' two-tailed t-test. Data analysis was processed by Origin data analysis software. Statistically significant criteria are: * $p < 0.05$, ** $p < 0.01$, *** $p < 0.001$ as significant differences.

2.5 Experimental Results and Discussion

2.5.1 Phage Display Technology Selects Specific Binding Phage REOB-1

Phage display technology was used to screen phages that specifically bound to nano-Nd_2O_3. After mixing the phage library with Nd_2O_3 nanosuspension, shaking and incubating in a 37° constant temperature shaker to fully bind the phage to the library. After centrifugation at high speed, unbound phages were washed away, and phages bound to the surface of Nd_2O_3 nanoparticles were recovered. In this way (combination-recovery-amplification-combination) for three rounds of screening, a large number of phages not bound to the nanoparticles were screened out, and phages specifically bound to the nanoparticles were fully enriched. About 100 phages were recovered at the end of the first round of screening. After the third round of amplification, the binding efficiency of phages was improved by two orders of magnitude compared to the phage pool. We randomly picked 15 monoclonal bacteriophages from the third round of recovered biocides. Each picked monoclonal was amplified separately, and phages were extracted for sequencing. Sequencing results revealed that 12 of the 15 monoclonal phages had identical sequences, and the insertion sequences of their surface display peptides were all TARSPWI. Therefore, we named this highly specific phage as REOB-1, and the encoded peptide sequence was CTARSPWIC (Fig. 2.1).

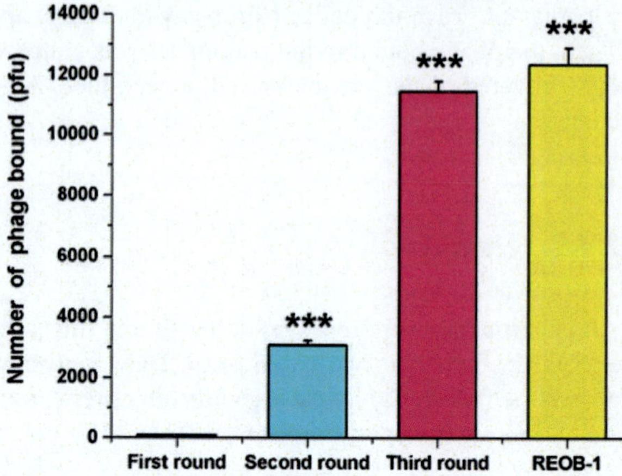

Fig. 2.1 Three rounds of screening of REOB-1 phages by phage display technology

2.5.2 *Identification of Phage Binding by PCR*

In order to prove the unique effect of phage REOB-1, we randomly selected a random phage from the phage library as a control, named it AP-1, and its encoded peptide sequence was CNATLPHQC. Mix the REOB-1 phage and AP-1 phage with a 1000-fold titer difference, and add the mixed phages to the nano-Nd_2O_3 particles at the same time, so that the two phages compete to bind the nanoparticles under the same conditions. After multiple washings, wash away unbound phages and recover phage bound to the surface and spread on a plate. 50 blue plaques were randomly selected, each chew plaque was divided into two, and two PVR specific primers were used to perform a double PVR reaction simultaneously. The results showed that 48 of the 50 monoclonal phages were REOB-1 phages and the other 2 were AP-1 phages. Because the titers of the added phages differ by 1000 times, the binding force of REOB-1 phage to the nanoparticles is 24,000 times higher than that of AP-1 phage. As shown in Fig. 2.2, 12 of the 13 monoclonal phages were reob-1 phages, and only 1 was AP-1 phage. It shows the highly specific and high affinity binding ability of REOB-1 phage to nanoparticles. At the same time, we tried to identify the binding efficiency of various rare earth metal oxides and rare earth up-conversion luminescent nanomaterials with this specific phage, and found that REOB-1 phages also bind to these rare earth nanomaterials with high affinity (Fig. 2.2).

2.5.3 *TEM Observation of Binding Form of REOB-1 Phage to UCN*

The high affinity phage that specifically binds rare earth nanomaterials was obtained through screening, but how did the phage and rare earth nanomaterials combine? It

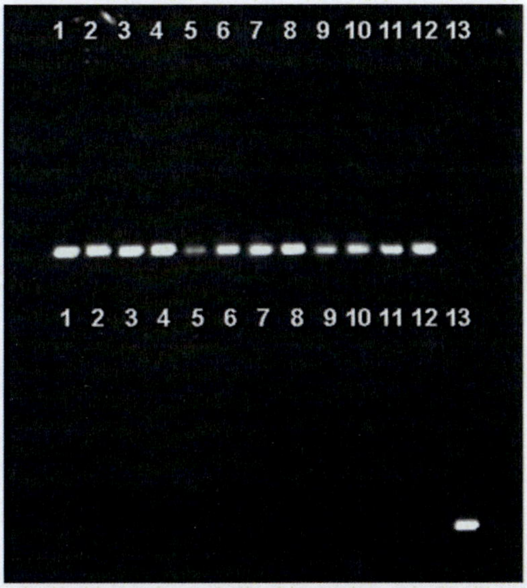

Fig. 2.2 Identification of binding capacity of REOB-1 phages by PCR

has been reported that TEM is used to observe the combination of phage and nanomaterials. In order to observe the combined form of phage and rare earth nanomaterials more visually, and the microscopic morphology of phages bound to the surface of rare earth nanomaterials, we have selected regular, approximate the spherical rare earth up-conversion luminescent nano-material UCN to observe by combining with the observation of a transmission electron microscope. According to the method, UCN, REOB-1 phage, and the binding products of REOB-1 phage wrapped on the surface of UCN were observed. As shown in the figure, under the transmission electron microscope, it can be clearly seen that the unwrapped UCN is approximately disc-shaped with a size of less than 100 nm. After staining, the REOB-1 bacteriophages were filamentous, of about 1 um in length and several nanometers in diameter. In the sample group of phage-UCN binding products, a large number of phages were tightly bound to the surface of the spherical nanoparticles, with one end of the phage adheres to the surface of the UCN, while the other end is distributed "radially" around the UCN (Fig. 2.3).

Unencapsulated UCN is shown on the left, REOB-1 phage is shown in the middle, and REOB-1 phage is shown binding to UCN on the right. The scale bar: 100 nm.

2.5.4 RE-1 Inhibits the Binding of REOB-1 to Nano-Nd_2O_3

By observing the binding form of REOB-1 phage to UCN, we found that REOB-1 phage was tightly bound to the surface of UCN with one end. According to the

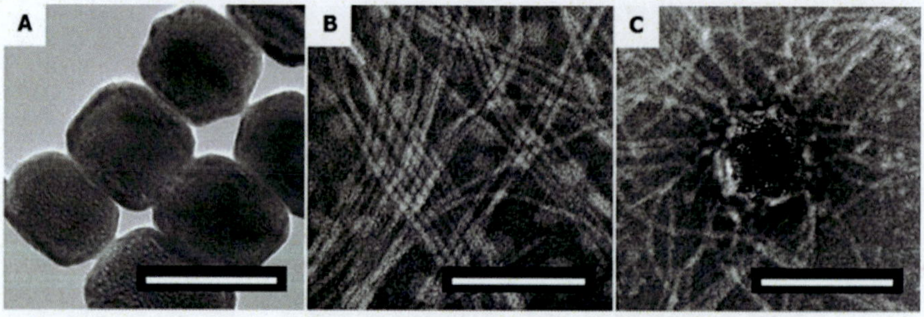

Fig. 2.3 Transmission electron microscopy image showing binding of REOB-1 phage to UCN

structure of the bacteriophage, the N-terminus of the bacteriophage is the P III protein displaying the surface polypeptide, so we speculate that the REOB-1 bacteriophage is tightly bound to the rare earth nanomaterial through the surface displayed polypeptide. Therefore, we synthesized RE-1 according to the sequence encoded by the phage. The amino acids A and G at ends of the polypeptide derived from the capsid protein of REOB-1 phage, and are used to further enhance the stability of the polypeptide.

In order to further verify whether the REOB-1 phage is tightly bound to the rare earth nanomaterial through the surface display peptide RE-1, we designed a short peptide and phage-nano competitive binding experiment, basing on the RE-1 short peptide not only possessing the same sequence as the phage, but also having a molecular weight of two orders of magnitude less than that of the phage. In the absence of the short peptide RE-1, about 12,500 REOB-1 phages bind to nano-Nd_2O_3. The number of phages bound to nano-Nd_2O_3 in the presence of different concentrations of short peptides RE-1 was calculated respectively, and the inhibition rate of phage binding by RE-1 with concentration gradient was calculated. The statistical results show that as the concentration of RE-1 increases, the number of phage bound to the nanometer decreases; RE-1 will inhibit the binding ability of the phage to the nanometer Nd_2O_3 as the dose increases; When the concentration reached 50 ug/ml, the binding ability of REOB-1 phage and Nd_2O_3 was almost completely inhibited. In the control group, even 5 mg/ml of the control phage AP-1 was used to compete for the binding of REOB-1 phage and nanometer Nd_2O_3. The proportion of suppressed REOB-1 phage was is still less than 20%. This result indicates that the short peptide RE-1 has a competitive inhibitory effect on REOB-1 bacteriophages, and proves that the high affinity binding of REOB-1 bacteriophages to rare earth nanomaterials is due to the specificity interaction between the polypeptide RE-1 displayed on the surface and the surface of the nanomaterials (Fig. 2.4).

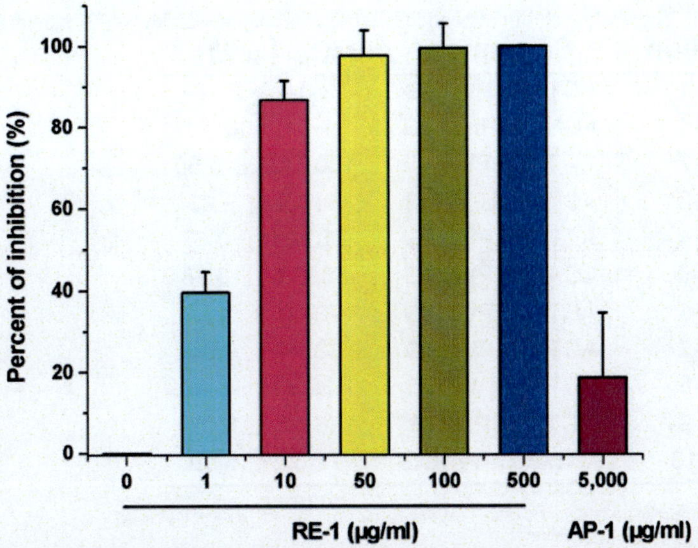

Fig. 2.4 Effects of Re-1 peptides of different concentrations on the competition of REOB-1 phages

2.5.5 Binding of Peptide Analogs to Nano-Nd_2O_3

In the above experiments, we found that the specific binding REOB-1 phage by phage display technology achieved high affinity binding with rare earth nanomaterials through the peptide RE-1 displayed on its surface. The experimental results of the control peptide set in the peptide competition experiment suggested that the short peptide RE-1 bound to the nano-Nd_2O_3 has sequence specificity. To further verify this conjecture, we synthesized a series of short peptide analogs. Using method 2.2.3.7, the competitive inhibitory ability of the peptide analogues on the binding of REOB-1 phage to nano-Nd_2O_3 was tested. The results are shown in Table 2.1. Polypeptide RE-1 and its analogs have different degrees of binding power to nano-Nd_2O_3. The IC_{50} value reflects the relative activity of various peptides to competitively inhibit the binding of REOB-1 phage to nano-Nd_2O_3. The smaller the IC_{50} value is, the stronger the ability of the peptide to bind to Nd_2O_3 nanometers is. Further analysis of the results showed that the short peptide RE-2, which possesses only seven amino acids in the middle of RE-1, bound 400 times worse than RE-1, while binding capacity ratio of the RE-4 (first cysteine changed to propyl) was also 90 times worse. These indicated that the two cysteine residues on both sides of RE-1 are extremely important for the binding of RE-1 short peptide to nano-Nd_2O_3. RE-3 has the same amino acid composition, but only disrupts the order of the 11 internal amino acids. Its IC_{50} value is only 20 times that of RE-1, whose binding ability is just worse than RE-1 among all the peptide analogs. This result is consistent with the viewpoints of previous articles published by Goede et al. [1], both of which prove that in the selection of inorganic nanomaterials-binding peptides, the amino acid sequence of the peptide but not the

Table 2.1 The binding ability of peptide RE-1 and its analogues with Nd_2O_3 nanomaterials

Peptide	Sequence	IC_{50} (μM) *
RE-1	ACTARSPWICG	3.01 ± 0.14
AP-1	ACNATLPHQCG	>5,000
RE-2	TARSPWI	1307.14 ± 6.50
RE-3	ACWPATRISCG	64.29 ± 2.33
RE-4	AATARSPWICG	270.04 ± 8.67
RE-5	ACAARSPWICG	688.70 ± 8.38
RE-6	ACTAASPWICG	78.41 ± 7.07
RE-7	ACTARAPWICG	83.94 ± 6.65
RE-8	ACTARSAWIGC	439.61 ± 8.48
RE-9	ACTARSPAICG	86.67 ± 8.07
RE-10	ACTARSPWACG	71.52 ± 5.10

sequence of amino acids that make up the peptide plays a key role. For the other six short peptides of RE-5 to RE-10, they are characterized in that a single amino acid in the middle 7 amino acids is sequentially replaced with alanine from left to right. By analyzing the IC50 value, it can be found that among the 7 amino acid residues in the middle of the RE-1 short peptide, threonine and proline are more important than the remaining amino acid residues.

2.6 Summary

In this chapter, we successfully used phage display technology to select phage RE-1 that specifically binds to rare earth nanomaterials. It was found that the REOB-1 phage specifically interacted with rare earth nanomaterials through the short peptide RE-1 displayed on its surface, achieving its high affinity binding. By verifying the binding ability of a series of variant peptides of RE-1 with rare earth nanomaterials, we further found that the binding of short peptide RE-1 to rare earth nanomaterials has sequence specificity, and that two cysteine residues at both sides of RE-1 are extremely important for the binding of RE-1 to nano-Nd_2O_3. Among the 7 amino acid residues in the middle of the RE-1 short peptide, threonine and proline are more important than the other amino acid residues. We have also shown that in the selection of inorganic nanomaterial-binding peptides, the amino acid sequence of the peptide plays a key role, but not the sequence of amino acids that make up the peptide.

However, the specific binding of phages selected by phage display technology to rare earth nanomaterials does not necessarily mean that synthesized short peptides chemically can be bound. To this end, we further verified the binding ability of short peptides to rare earth nanomaterials.

Reference

1. Goede, K., Busch, P., & Grundmann, M. (2004). Binding specificity of a peptide on semiconductor surfaces. *Nano Letters, 4*(11), 2115–2120.

Chapter 3
Binding Rare Earth Nano Materials with High Affinity and Forming Surface Coating

3.1 Introduction

In the previous chapter, it was proved that the phage REOB-1 with specific and high affinity binding to rare earth nanomaterials was screened by phage display technology. The binding of peptides on the surface with rare earth nanomaterials is sequence specific. However, many previous reports have shown that phage physical binding does not necessarily mean it can chemically bind synthesized peptides, so it is important to show that peptides can efficiently bind nanomaterials. In this chapter, we have carried out a series of validation experiments to identify the binding ability of RE-1 short peptide and rare earth nano materials. The properties of RE-1 nano materials before and after RE-1 coating were compared. The binding ability of RE-1 short peptide to rare earth nanomaterials was verified from the aspects of material, peptide, peptide conformation and interaction. At the same time, there are few electron micrographs in peptide reports, and we have successfully realized the intuitive observation of the binding form of peptide and nanomaterials.

3.2 Experimental Materials

3.2.1 Reagents

·YCl_3, $YbCl_3$, $Ercl_3$, NH_4F, $Si0_2$, Si0, Nano silver, diamond, TiO_2, Y_2O_3, phosphotungstic acid were all purchased with sigma company.

- UCP(PTIR550/F,$NaY_{0.77}Yb_{0.21}Er_{0.02}F4$) were purchased from stevnich, UK Company.
- Methanol, ethanol, cyclohexane, oleic acid, octadecene, NaOH, concentrated nitric acid, etc. are purchased from the chemical reagent Library of University of science and technology of China.

Y. Zhang, *Tuning Autophagy-Inducing Activity and Toxicity for Lanthanide Nanocrystals*, Springer Theses, https://doi.org/10.1007/978-981-16-8166-0_3

- APS (F201 10,517, Sinopharm Chemical Reagent Co., Ltd., China).
- Glycerin (sjll21s8011z, biotechnology, China).
- SDS (107k0110, sigma, USA).
- Tris (mc0718820012j, biotechnology, China).
- Tricine (mc041181012j, biotechnology, China).
- Acrylamide (mc0426810012j, biotechnology, China).
- TEMED (0100c327, Amresco, USA).
- 50 mm Tris·HCL (pH 7.0) buffer: prepare 1 M Tris· HCL(pH 7.0), weigh 121 g Tris, add 800 ml H_20, dissolve it, then adjust the pH value to 7.0 with concentrated hydrochloric acid, and then fix the volume to 1L. When used, dilute 1 M Tris·HCL (pH 7.0) to 50 mM.
- 10% SDS: weigh 1 g SDS, add H_20 to constant volume to 10 ml, for standby.
- 2% phosphotungstic acid (pH 6.7): weigh 1 g of phosphotungstic acid (PTA) powder and prepare it into a concentration of 2%
- 1 mol/L Potassium Hydroxide Adjusted the pH to 6.7 and Stored at 4 °C for Standby

3.2.2 Experimental Equipment and Consumables

- Special sample cell for particle size analysis (dts0012, malvem, UK).
- Special sample cell for potential analysis (dts0013, malvem, UK).
- Fluorescent cuvette (×72,052, alpha, UK).
- UV detection cell (×72,053, alpha, UK).
- Protein gel electrophoresis tank (552br / 032,042, bio. Rad, USA).
- Long neck bottle, copper mesh sample box and other consumables are purchased from the chemical reagent Library of University of science and technology of China.Special sample cell for particle size analysis (dts0012, malvem, UK).
- Special sample cell for potential analysis (dts0013, malvem, UK).
- Fluorescent cuvette (×72,052, alpha, UK).
- UV detection cell (×72,053, alpha, UK).
- Protein gel electrophoresis tank (552 br/032,042, bio. Rad, USA).
- Long neck bottle, copper mesh sample box and other consumables are purchased from the chemical reagent Library of University of science and technology of China.

3.3 Instrument and Equipment

Vortex mixer (MSL minishake, IKA, China).

- ·Ultrasonic cell breaker (jy92.11, Nanjing xianou Instrument Manufacturing Co., Ltd., China).

- Inductively coupled plasma mass spectrometry (Agilent 7500 CE, Agilent Technologies, USA).
- Milli. Q ultra pure water purification system (direct-Q 3, millipore, USA).
- Fluorescence spectrophotometer (RF-5301PC, Shimadzu, Japan).
- UV spectrophotometer (DU-640, Beckman, USA).
- 980 nm infrared laser (MDL·980nm, 1 W, Changchun new industry photoelectric technology company, China).
- Malvem Laser Particle Sizer (nano zs90, malvern, UK).
- Transmission electron microscope (JEM. 2100f, jeol, Japan).
- Scanning electron microscope (JSM. 6700f, jeol, Japan).
- Nuclear magnetic resonance spectrometer (avance 500.132 MHz, Bruker, Germany).
- Isothermal titration microcalorimeter (VP-ITC, Microcal, USA).
- Biomolecular interaction analyzer (Biacore 3000, Ge, USA).
- Circular dichroic spectrometer (J-810, JASCO, Japan).
- Fourier infrared spectrometer (IFS 125 h, Bruker, Germany).

3.4 Methods

3.4.1 Preparation of Rare Earth up Conversion Luminescent Nanomaterials

3.4.1.1 Synthesis of UCN/UCN-S ($NaYF_4$: Yb, Er)

The rare earth up conversion luminescent nanomaterials (UCN and UCN-S) used in this paper are spherical $NaYF_4$:18% Yb, 2% Er nanoparticles were synthesized as follows: YCl_3(0.1562 g), YbC13(0.0503 g),ErC13 (0. 0055 g) mixed with 6 ml oleic acid and 17 ml octadecene in a 50 ml long neck flask, heated to 160 °C, then cool to room temperature. Dissolve NaOH (0.01 g) and NH_4F (0.148 g) into 10 ml Methanol Add to the long neck bottle one by one and stir 30 rain to ensure the chloride reacts fully. Put the above solution slowly heat and evaporate methanol, Degas at 100 °C for 10 min, then heat to 300 °C in argon and maintain 1 h. When the solution cools naturally, the nanocrystals precipitate. Precipitate with ethanol, ethanol / water (volume in 1:1 ratio), After washing three times the oil on the surface of nanoparticles was removed by treating with 1 M HCL for 5 h at room temperature, wash with water and cyclohexane for 10 times. Detection of 230 nm (characteristic absorption of oleic acid) by UV spectrophotometer. To ensure that the oleic acid is completely removed. In the synthesis process, different amounts of oil are added to control the particle size of UCN and UCN-S nanoparticles (UCN: 3 ml oleic acid, UCN-S 6 ml Oleic acid. The synthesized UCN is stored in water for use in experiments. (Note: due to the removal of UCN and the oleic acid on the surface of

UCN-S nanoparticles is in the form of naked dew, extremely unstable and easy to agglomerate, Therefore, nanoparticles need to be prepared on demand).

3.4.1.2 Pretreatment Methods of UCP

UCP (PTIR550/F, $NaY_{0.77}Yb_{0.21}Er_{0.02}F_4$) UCP was purchased from the fluorescence technology company of stevnich, UK, with an average particle size of about 3 μM. Due to the particle size of the up-conversion fluorescent material reaches the micron level, so it needs to be pretreated to nanometer level before it can be used in experiments. UCP was suspended in water to prepare a 10 mg/ml stock solution, centrifuged at 500 rpm for 10 min to discard the precipitate to remove the large particles, the supernatant was transferred into a 50 ml centrifuge tube, placed in a 120 W ultrasonic cell breaker for 1 h, and centrifuged at 500 rpm for 10 min. The above process was repeated three times to obtain UCP with an average particle size of about 500 nm. The UCP obtained by the above method was analyzed by ICP-MS (see method 3.4.2 for specific experimental method), and then suspended in water to prepare a 5 mg / ml stock solution for further experiments.

3.4.2 Determination of Rare Earth Up-Conversion Luminescent Nanomaterials by ICP-MS

The concentrations of UCP, UCN and UCN-S were determined by ICP-Ms. Take 50μL of synthesized or pretreated UCP, UCN and UCN-S nanoparticles respectively and dissolve them in 300 μL of concentrated nitric acid solution. After 2 h, until the nanoparticles are completely melted, the laser with 980 nm excitation light can be used for detection. If there is still green fluorescence in the solution, it means that the particles are not completely dissolved. After the nanoparticles are completely dissolved, they can be used for ICP-MS analysis.

3.4.3 Test the Binding Ability of RE-1 with Various Nano Materials

3.4.3.1 Establishment of FITC-RE-1 Fluorescence Value Concentration Relationship Standard Curve

Six different concentration gradients of FITC-RE-1 polypeptide were diluted respectively, and the peak value of 525 nm of each gradients was detected by 488 nm excitation spectrum in the fluorescence spectrophotometer, and the standard curve was drawn by the relationship between the fluorescence value and concentration of

FITC-RE-1. By fitting the standard curve, the relation equation between fluorescence value x and concentration y is: y = 0.0008x-0.0007 ($R^2 = 0.9976$).

3.4.3.2 Test Method of Combining Ability

The 10 μg FITC-RE-1 polypeptide was mixed with more than 100 μL inorganic nano materials (1 mg/ml, such as nano SiO_2 /Si03, nano silver, diamond, TiO_2, UCN, UCP, UCN-S, Y_2O_3 and other nano rare earth metal oxides) and incubated at room temperature for 1 h, then centrifugate at 12,000 rpm for 10 min, the precipitated nanoparticles were washed with milli-Q water for three times, and FITC (excitation wavelength 488 nm, emission wavelength 535 nm) was detected by fluorescence spectrophotometer. The fluorescence value of FITC-RE-1 was used to determine the relationship between concentration and fluorescence value. According to the standard curve equation, the binding ability of different materials with FITC-Re-1 polypeptide was obtained.

Note: the newly prepared material should be used after ultrasound before binding with polypeptide. FITC-RE-1 is the fluorescent molecule FITC coupled to the amino terminal of RE-1 short peptide through ACP.

3.4.4 RE-1 Combined with Rare Earth Up-Conversion Luminescent Nanomaterials

Add 50 μg RE-1 or fluorescence peptide FITC-RE-1 to 100 μL UCN (1 mg/ml), 10 μg RE-1 or fluorescence peptide FITC-RE-1 to 100 μg UCP (1 mg/ml), 300 μg RE-l or fluorescence peptide FITC-RE-1 to 100 μg UCN-S (1 mg/ml), mix them evenly, shake them gently at room temperature, incubate for 1 h. After centrifugation at 12,000 rpm for 10 min, the precipitated nanoparticles were washed with milli-Q water for 2–3 times, and then resuspended with appropriate water or solvent according to the type of experiment for different experiments.

3.4.5 Identification Method of Binding Ability of RE-1 and Rare Earth Up-Conversion Luminescent Nanomaterials

3.4.5.1 Identification of Combination Time

The binding ability of RE-1 short peptide to up-conversion luminescent nanomaterials (UCN, UCP, UCN-S) was tested. In accordance with method 3.4.4, RE-1 peptide was incubated with UCN, UCP and UCN-S at room temperature by gently shaking different time (2, 10, 30, 120, 720 min).

After centrifuging the binding products at different time points in a centrifuge at 12,000 rpm for 10 min, the precipitated nanoparticles were washed with milli-Q water for 2–3 times, and finally re suspended in 100 μL of water. After ultrasonic treatment for 1 min, according to method 3.4.3.2, the amount of FITC-RE-1 fluorescein bound on the surface of UCN was detected by fluorescence spectrophotometer.

3.4.5.2 Identification of Binding Concentration

After mixing 100 μg UCN with FITC-RE-1 of different concentrations (5, 10, 25, 50, 100 μg), the precipitated nanoparticles were washed with milli-Q water for 2–3 times, wash off the excess unconjugated peptide, and then suspend it in 100μL water again. Then, ultrasound for 1 min. according to method 3.4.3.2, use fluorescence spectrophotometer to detect the amount of FITC-RE-1 fluorescent peptide on the surface of UCN. The binding concentration of RE-1 short peptide with UCP and UCN- S was detected by the same method.

Due to the different specific surface areas of UCN, UCP and UCN-S, the amount of peptides bound to different materials is also different. Therefore, FITC-Re-1 combined with 100 μg UCP was tested at concentrations of 1, 5, 10 and 20 μg. The concentration gradient of 100 μg FITC-RE-1 combined with 50, 100, 200, 300, 400 μg UCN-S.

3.4.6 Experimental Method of AP-1 and RE-1 Competitive Rare Earth Up-Conversion Luminescent Nano Materials Combined with UCN

50 μg FITC-RE-1 and 10 μg AP-l were added to 100 μg UCN suspension at the same time. The control group was 50 μg FITC-RE-1 without AP-1. The mixture of 1 and 100 μg UCN was mixed evenly in two tubes of suspension, incubated at room temperature for 1 h, centrifuged at 12,000 rpm for 10 min, and the precipitated UCN nanoparticles were washed with milli-Q water for 2–3 times, remove the excess unbound peptide, and finally suspend it in 100μL water again, and then conduct ultrasound for 1 min. according to method 3.4.3.2, use fluorescence spectrophotometer to detect the amount of FITC-Re-1 fluorescent peptide bound on the surface of UCN. For large particle UCP, except the amount of FITC-Re-1 peptide, the other methods was same, then add 100 μg UCP.

3.4.7 *Effect of Solution Environment on the Binding Ability of RE-L and Rare Earth Up-Conversion Luminescent Nanomaterials*

The influence of solution environment (pH, salt ion concentration, temperature) on the binding ability of RE-1 short peptide and rare earth up-conversion luminescent nanomaterials (UCN, UCP) was detected. The influence of the environment on the binding capacity of rare earth up-conversion luminescent nanomaterials was judged by measuring the amount of binding fluorescent peptides on the surface of the materials in different solution environments.

3.4.7.1 Effect of Different pH Values on Binding Capacity

According to method 3.4.4, 50 μg FITC-Re-1 and 10 μg FITC-RE-1 were respectively dissolved in different Ph (3, 5, 7, 10).After mixed with 100 μg UCN and UCP, the incubation is conducted at room temperature of 25 °C by gently shaking. After centrifugation at 12,000 rpm for 10 rain, the precipitated nanoparticles were washed 2–3 times with milli-Q water and washed away the unconjugated pepetide. Then suspend it again in 100 μL water, ultrasound gently for 1 min, and detect the amount of FITC-Re-1 fluorescent peptide on the surface of UCN with fluorescence spectrophotometer according to method 3.4.3.2.

3.4.7.2 Experiment on the Influence of Different Salt Ion Concentration on the Binding Capacity

According to method 3.4.4, 50 μg FITC-RE-1 and 10 μg FITC-RE-1 were mixed with 100 μg UCN and UCP in different salt ion concentrations (0,0.15, 0.3, 0.5 M NaCl), and incubated at 25 °C for 1 h. The precipitated nanoparticles were washed with milli-Q water for 2–3 times, and the excess nonbinding peptides were removed. Finally, they were suspended in 100 μL water for 1 min, and the amount of FITC-RE-1 binding peptides on UCN surface was detected by fluorescence spectrophotometer according to method 3.4.3.2.

3.4.7.3 Experiment on the Influence of Different Temperatures on the Binding Capacity

According to method 3.4.4, 50 μg FITC-Re-1 and 1 μg FITC-RE-1 were mixed with 100 μg UCN and UCP dissolved in milli-Q water, respectively, and then incubated in different temperature (25, 37, 55, 80 °C) for 1 h, and centrifuged at 12,000 rpm for 10 min. the precipitated nanoparticles were washed with milli-Q water for 2–3 times, and the excess unconjugated peptides were removed. Finally, they were resuspended in milli-Q water. The amount of FITC-Re-1 fluorescent peptide on the surface of UCN was detected by fluorescence spectrophotometer according to method 3.4.3.2.

3.4.8 *Dissociation of RE-1 and Rare Earth Up-Conversion Luminescent Nanomaterials (UCN, UCP)*

3.4.8.1 Fluorescence Spectrophotometer Detection Method

According to method 3.4.4, 250 μg FITC-RE-1 was fully combined with 500 μg UCN, and the excess unbound polypeptide was washed away. Then, the binding product was suspended in 500 μL water again, followed by a brief and gentle ultrasound for 1 min, and evenly divided into five equal parts. The product was gently shaken on the shaking bed at room temperature, and centrifuged at different time points (0, 2, 12, 24, 48 h) for 10 min at 12, 000 rpm respectively, the amount of FITC-RE-1 separated from the supernatant was detected by fluorescence spectrophotometer.

3.4.8.2 Tricine SDS Small Peptide Gel Preparation Method

Anode buffer (1×): 0.2 M Tris (pH 8.9): 12.114 g, add H_20 to a constant volume of 500 ml, and store at 4 °C.

Cathode buffer (5×): configure 0.1 M Tris, 0.1 M Tricine and 0.1% SDS. 30.285 g Tris, 44.8 g Tricine, 2.5 g SDS and H_20 were added to fix the volume to 500 ml and stored at 4 °C.

Gel buffer (G-B, 3×): 36.342 g Tris, 0.3 g SDS, add H_20 to constant volume to 100 ml, adjust the pH value of HCL to 8.45, filter and store at 4 °C.

AB-3 storage solution (49.5% T, 3% C mixture): 24 g of acrylamide, 0.75 g of methylbisacrylamide, add H_2o to a constant volume of 50 ml, filter and store at 4 °C.

AB-6 storage solution (49.5% T, 6% C mixture): 23.25 g of acrylamide, 1.5 g of methylbisacrylamide, add H_2o to a constant volume of 50 ml, filter, and store at 4 °C.

	4% laminated adhesive	10% separator
AB-3 (mL)	0.25	–
AB-6 (mL)	–	1.6
3× gel buffer (mL)	0.75	2.67
3× gel buffer (mL)	–	1.26
H_2O (mL)	2	2.47
10% APS (μL)	22.5	40
TEMED (μL)	2.25	4

3.4.8.3 Tricine-SDS Electrophoresis

According to the combination method of method 3.4.4, 40 μg FITC-Re-1 was fully combined with 400 μg UCP, and the excess unbound polypeptide was washed away. Then the binding product was suspended in 400 μL water for a short period of gentle ultrasound for 1 min, and evenly divided into four equal parts and put into four centrifuges. After centrifugation at 12,000 rpm for 10 min, the supernatant was discarded, and the precipitates (binding products) in four tubes were treated as follows:

The first tube was resuspended with 50 mM Tris–HCL (pH 7.0) buffer, and ultrasound was performed for 1 min. After that, the rats were incubated at room temperature and shaking for 2 h;

The second tube was resuspended with glycine. HCl (pH 2.0) buffer solution, and ultrasound was performed for 1 min, incubate at room temperature and shaking for 2 h;

The third tube binding product was resuspended with 50 mM Tris-HCI (pH 7. 0) buffer, and ultrasound was performed for 1 min. After that, the water bath was heated and boiled for 10 min;

The fourth tube was resuspended with 10% SDS, and incubated with light ultrasound for 1 min at room temperature and gentle vibration for 2 h.

After treatment, centrifugation was performed at 12,000 rpm for 10 min, and all supernatants were taken for measurement.

10%Tricine-SDS small peptide glue was prepared, and all the samples were run on electrophoresis. The gel imaging system was used to record the results.

3.4.9 Calculation Method of Binding Molecular Number of RE-1 and Rare Earth Up-Conversion Luminescent Nanomaterials

Assuming that UCN and UCN-S are approximately spheres, the average weight W of each nanocrystal can be calculated as follows:

$$W = \left[4/3 \times \pi R^3\right] \times \rho$$

In this formula, R represents the radius of the nanocrystal (UCN is about 46 nm, UCN- S is about 10 nm), ρ is the density in the lattice of β-$NaYF_4$ nanocrystals (according to 16.0334 recorded in the JCPDS card database of powder diffraction, the density ρ of β-$NaYF_4$ is 4.309 g/cm^3).

According to the known surface area of nanocrystals (S) $= 4\pi R^3$, NA is the Avogadro constant (3.023×10^{23}/mol). The following conclusions can be drawn:

The number of molecules per 100 μg of UCN nanocrystals is 5.69×10^{10}
The number of molecules per 100 μg of UCN-S nanocrystals is 5.54×10^{12}

The surface area of a single UCN nanocrystal is 2.66×10^4 nm^2
The surface area of a single UCN-S nanocrystal is 1.256×10^3 nm^2

Therefore, the total surface area of 100 μg UCN nanocrystals is 1.51×10^{15}nm^2, the total surface area of 100 μg UCN-S nanocrystals is 6.96×10^{10}nm^2. According to the calculation, a single UCN nanoparticle can combine 3.2×10^5 FITC-Re-1 peptide molecules; a single UCN-S nanoparticle can combine 2.1×10^4 FITC-Re-1 peptide molecules. It is known that the molecular weight of fluorescence peptide FITC-Re-1 is 1553.8 Da, so the maximum number of FITC-Re-1 peptides that can be combined is 12.03 per square nanometer of UCN nanoparticles, while the maximum number of FITC-Re-1 peptides that can be combined is 16.77 per square nanometer of UCN-S nanoparticles.

3.4.10 Identification of Nanoscale and Potential of Rare Earth Up-Conversion Luminescent Nanomaterials

3.4.10.1 DLS Particle Size Analysis

DLS was used to detect the particle size of up-conversion luminescent nano materials (UCN, UCP, UCN-S). Using method 3.4.4, the RE-1 short peptide was fully combined with UCN, and the excess short peptide was washed away. The binding product was resuspended with 1 mL milli-Q water. Take the same amount of UCN, UCP and UCN-S suspension which have been synthesized and pretreated, and fix the volume to 1 mL with milli-Q water. After the above-mentioned nano materials coated with re.1 and uncoated are prepared and tested by light ultrasound for 1 min, they are respectively added into a special sample cell for particle size analysis by using a pipette gun. The particle size distribution is measured by Malvern nano laser particle size analyzer with 633 nm He/Ne laser, and the results are recorded.

3.4.10.2 δ-potential Analysis

According to method 3.4.10.1. After the coating of RE-1 and UCN for 1 min, the samples were added to the special sample cell for potential analysis with electrodes. The potential of the samples was measured by the Malvern nano laser particle size analyzer with 633 nm He/Ne laser, and the charged condition of the nano particles was analyzed.

3.4.11 *Competitive Experiment Method of Peptide and Peptide*

Prepare 5 tubes of mixed suspensions of 200 μg RE-5 and 100 μg UCN and add FITC-RE-1 with different concentration gradients (5 μg, 10 μg, 20 μg, 50 μg, 100 μg) to each tube respectively; prepare 5 tubes of mixed suspensions of 200 lag re.1 and 100 Xu UCN for n sample, and add FITC-RE-1 with different concentration gradients (5 μg, 10 μg, 20 μg, 50 μg, 100 μg) to each tube respectively. The above-mentioned 10 tube mixture was evenly mixed, incubated at room temperature with a slight vibration for 1 h, centrifuged at 12,000 rpm for 10 min, and the precipitated UCN nano meter.The particles were washed with milli-Q water for 2–3 times, and the excess unbound peptides were removed. Finally, the particles were suspended in 100 L of water again, and ultrasound was performed for 1 min. according to method 3.4.3, 2,fluorescence spectrophotometer was used to detect the amount of FITC-Re-1 fluorescent peptides bound on the surface of UCN.

3.4.12 *TEM Observation of the Combination Form of RE-1 and UCN*

Transmission electron microscopy (TEM) was used to observe the binding forms of RE-1 peptide and UCN. According to method 3.4.4, RE-1 short peptide (high concentration of 5 μg, low concentration of 50 μg) was combined with 100 μg UCN respectively, then the redundant and unbound peptide was washed away, and the binding product was stained with 2% phosphotungstic acid (pH 6.7).

The copper mesh coated with carbon film is clamped on the filter paper with special tweezers for electron microscope, the reflective side of the copper mesh is facing down, and the micro pipette gun is used to suck 1μL. The above dyed sample suspension is slightly dripped onto the non-reflective side of the copper mesh. After the sample is evenly distributed on the copper mesh, take a clean glass dish and cover the copper mesh to avoid dust pollution in the air. After the sample is dried naturally, the tem of 200 kV transmission electron microscope is used to observe the image and record the results. The observed copper mesh is placed in the copper mesh sample box for preservation.

Note: due to the high magnification of transmission electron microscope, the concentration of UCN should be appropriate. If the concentration is too low, the sample may not be found in the electron microscope field of vision; if the concentration is too high, the sample will be highly agglomerated and dense in the field of vision, and it is not easy to see a single nano UNC particle. Therefore, the above-mentioned binding products can be measured and observed after being slightly different multiple.

3.4.13 SEM Observation of the Combination Form of RE-1 and UCN

Scanning electron microscopy (SEM) was used to observe the binding form of re.1 short peptide and UCN. According to method 3.4.4, combine the RE-1 short peptide with UCN and wash away the redundant and unbound peptide. The cut silicon wafer with an area of about 4 mm × 4 mm is adhered to the sample of SEM by conductive tape.

On the stage, gently wipe the surface of the silicon wafer with the cotton swab dipped in the ethanol solution, suck 2μL of the above-mentioned binding products and the UCN suspension not covered with RE-1 into the surface of the single crystal silicon wafer with a micro pipette gun, and cover the copper mesh with a glass dish to avoid dust pollution in the air. After natural air drying of the sample, the images were observed and the results were recorded by scanning electron microscope (SEM) of 15 kV.

3.4.14 UV Vis Detection of the Binding of RE-1 to UCN

FITC-RE-1 combined with UCN surface was detected by UV VIS spectrophotometer. According to method 3.4.4, the FITC-RE-1 short peptide was fully combined with UCN, and the redundant and unbound peptide was washed away. The binding product was resuspended in 1 ml of milli-Q water, and added to the cuvette for ultraviolet visible 200–800 nm) full wavelength absorption spectrum scanning. At the same time, two control groups were set up, and the same amount of UCN and FITC-RE-1 were fixed to 1 ml with milli-Q water, respectively, for UV–Vis absorption spectrum detection.

3.4.15 ^1H NMR

3.4.15.1 Preparation of NMR Buffer

Prepare a buffer containing 20 mM Na_2HPO_4,100 mM NaCl (pH 7.0) and add 10% D_2O. D_2O can be added before detection.

3.4.15.2 Test Methods

According to method 3.4.4, fully combine 50 rugged RE-1 short peptide with 100 μg UCN, wash off the redundant and unbound peptide with NMR buffer, suspend the binding product in the above-mentioned NMR buffer again, add 10% D_2O into the NMR spectrometer for detection, and finally record the one-dimensional NMR hydrogen spectrum.

3.4.16 Fourier Transform Infrared Spectroscopy (FTIR)

According to method 3.4.4, 50 μg of RE-1 short peptide was fully combined with 100 μg of UCN, the redundant and unbound peptide was washed away, and the binding product was re suspended in 100μL of milli-Q water. After 30 min preheating by FTIR, milli-Q water drops were used as control in the liquid pool, and then UCN and RE-1 short peptide were dropped into the detection liquid pool (the thickness of the liquid film is generally 0.1–1 nm), respectively, and then put into the FTIR to detect, respectively record the scanning spectrum, and then analyze the spectrum after deducting the water peak of the control.

3.4.17 Circular Dichroism Spectroscopy (CD)

After the circular dichroism spectrometer is preheated, the milli-Q water is used as the control to scan the whole spectrum from 190 to 250 nm, and then 50 μg RE-1 short peptide is fully combined with 100 μg UCN, 50 μg RE-5 short peptide with 100 μg UCN, 50 μg RE-1 short peptide with 100 μg $Si0_3$ (without washing), according to method 3.4.4. The combined product is fully combined with unbound UCN, $Si0_3$, pure peptide RE-1 and pure peptide RE-5 under the same conditions. The circular dichroic spectrum scanning was carried out respectively, and each spectrum was the superposition result of three times of scanning.

3.4.18 Surface Plasmon Resonance (SPR)

50 μg Re-1 was covalently coupled to the sensor chip, and then 500μL UCN (including 100 μg) was added to the surface plasmon resonance (SPR) BIOCORE 3000 as the mobile phase. With the increase of the amount of UCN added, the change of the spectrum was recorded, and the spectral results were analyzed.

3.4.19 Isothermal Titration Microcalorimetry (ITC)

The binding capacity and interaction between RE-1 and UCN were studied by isothermal titration microcalorimeter. We used milli-Q water to dilute the concentration of RE-1 short peptide to 4 mM, and diluted the concentration of UCN nano material to 0.18 mM for full degassing and standby. The temperature of isothermal titration microcalorimeter is set at 25 ± 0.2 °C. Add the solvent (milli-Q water) for diluting RE-1 and UCN into the reference tank, and add the diluted 0.18 mM UCN into the sample tank. In order to prevent the bubble from affecting the experimental results, the mixture can be injected at the same time. Then add 4 mM of RE-1 short peptide diluted with milli-Q water into the syringe. Slowly insert the syringe into the sample cell and start titration. In the experiment, the stirring speed was 307 RPM / min, 13 drops were titrated, each drop was 3μL, the interval between the two drops was 120 s, and the titration stopped when the balance was reached. Blank experiment is the same setting.

Next, titrate the solvent milli-Q water with the same amount of RE-1 short peptide (add milli-Q water into the sample tank after the sample tank is cleaned). The heat change spectrum was recorded and the experimental data were processed and analyzed by the original 7.0 software package. According to the equation $\Delta G = -RT\ln KA$ and $\Delta G = \Delta H - T\Delta S$, the binding heat is obtained Mechanical constant.

3.4.20 Data Analysis

All experiments were repeated at least three times. All the data are represented by mean + S.E.M., and students are tested by two tailed t-test. Data analysis is processed by two tailed student's test in origin data analysis software. Statistically significant standard was $^*p < 0.05$, $^{**}P < 0.01$, $^{***}P < 0.005$ as the significant difference.

3.5 Results and Discussion

3.5.1 RE-1 Only Binds Rare Earth Nanomaterials Specifically

In the phage binding experiment in the previous chapter, we found that the REOB-1 phage can specifically bind to a variety of rare earth metal oxides and rare earth up-conversion luminescent materials. So, does the RE-1 short peptide bind to these rare earth nanomaterials? Does the re.1 short peptide bind to rare earth nanomaterials with material specificity? Based on these two questions, we tested a series of rare earth metal oxides (such as Y_20_3, Yb_2O_3, Er_2O_3, La_2O_3, CeO_2, Nd_2O_3, Eu_2O_3, Ho_2O_3, etc.) and non-rare earth nano materials (such as UCN, UCP, UCN-S, etc.) and non-rare earth nano materials (including metal oxides, such as TiO_2 metal nano materials, such as Ag; non-metal oxides, such as $Si0_2$, $Si0_3$; non The binding ability of metal elements such as diamond and RE-1 short peptide (see Sect. 3.4.4 for details). The results show that the fluorescent labeled RE-1 short peptide has strong binding

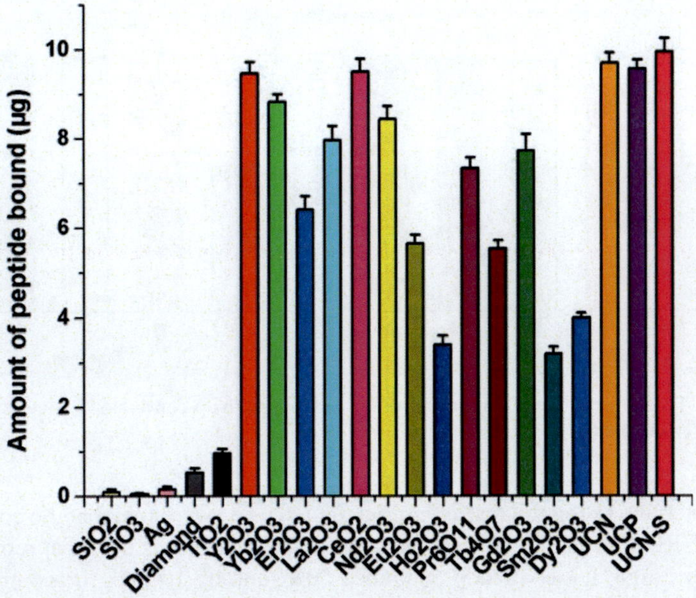

Fig. 3.1 Comparison between FITC-Re-1 and a variety of nanomaterials

capacity with rare earth nano materials, among which Y_2O_3, CeO_2, Yb_2O_3, Nd_2O_3, UCN, UCN-S (small particles of UCN) and UCP (large particles of UCN) have the strongest binding capacity. However, RE-1 and non-rare earth nano materials, such as diamond, TiO_2, SiO_2, SiO_3, and nano silver, are almost not combined (Fig. 3.1). The experimental results show that RE-1 short peptide can be widely combined with rare earth nanomaterials, but not with non-rare earth nanomaterials. It is confirmed that RE-1 short peptide is a kind of peptide specifically bound by rare earth nanomaterials, that is RE-1 has material specificity.

3.5.2 Determine the Binding Time of RE-1 and Rare Earth Up-Conversion Luminescent Nanomaterials

In the above experiments, we found that RE-1 short peptide specifically binds to rare earth nanomaterials, especially UCN, UCP, UCN-S and other rare earth up-conversion luminescent nanomaterials with high affinity. These rare earth up-conversion luminescent nanomaterials have unique up-conversion luminescent properties and rare earth element characteristics, which have great application potential in the biomedical field, making them stand out in many nanomaterials. Therefore, in the next experiment, we will focus on the research of this kind of materials. First of all, we studied the binding time of RE-1 and rare earth up-conversion luminescent

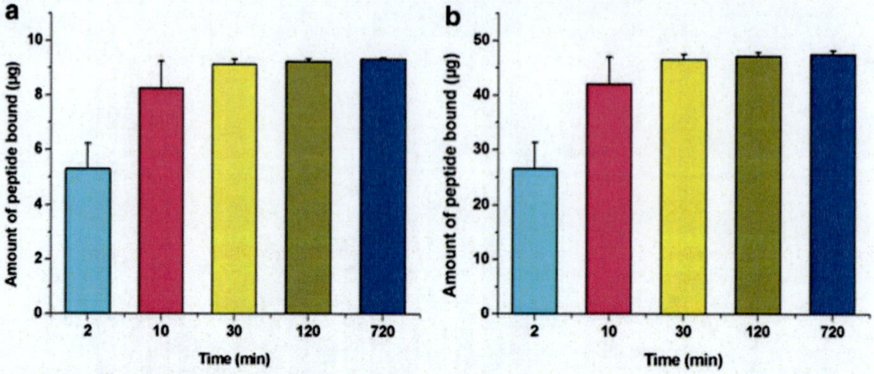

Fig. 3.2 Time effect of junction between RE-1 and rare earth up-conversion luminescent nanomaterials. **a** UCN, **b** UCP

nanomaterials. Using the research of method 3.4.5.1, we found that the combination of FITC-RE-1 and UCN is very rapid. When the platform is mixed at room temperature for 2 min, it has already combined 50%, and by 10 min, the combination of FITC-Re-1 and UCN can almost reach saturation (Fig. 3.2a). At the same time, for the large particle UCP, we also get the same conclusion of time effect in the combination experiment with FITC. Re-l (Fig. 3.2b).

3.5.3 *Determination of Binding Concentration of RE-1 and Rare Earth Up-Conversion Luminescent Nanomaterials*

Through the study of method 3.4.5.2, we further determined the binding concentration of RE-1 and rare earth up-conversion luminescent nanomaterials. After further analysis and calculation of the experimental results of this dose effect, we have come to the following conclusion: the amount of 100 µg UCN that can combine with FITC-RE-1 short peptide is 47 µg (about 30 nmol, the known molecular weight of FITC-RE-1 is 1553.8 DA) (as shown in Fig. 3.3a); The amount that 100 µg UCP can combine with FITC-RE-1 short belly is 9.2 µg (as shown in Fig. 3.3b): similarly, the amount of fitc-RE-1 combined with 100 µg UCN-S is 301 µg (as shown in Fig. 3.3c).

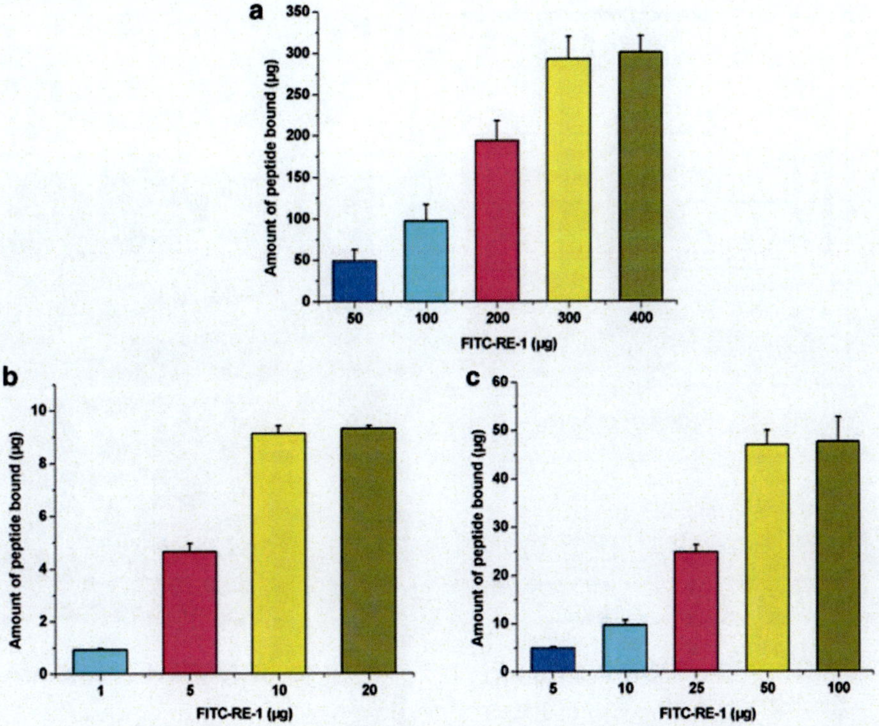

Fig. 3.3 Dose effect of RE-1 combined with rare earth up-conversion luminescent nanomaterials. **a** UCN, **b** UCP **c** UCN-S

3.5.4 Determination of the Size Distribution of Rare Earth Up-Conversion Luminescent Nanomaterials by DLS and TEM

In the above study, we found that although UCN, UCP and UCN-S have the same composition ($NaYF_4$: Yb, Er), and they are all rare earth up-conversion luminescent nanomaterials, but the amount of RE-1 short peptide that each nanomaterial combined is different, which may be caused by different specific surface areas of different materials, so we use dynamic light scattering (DLS) (see Sect. 3.4.10.1 for details). The nano scale of the three materials was detected, and the size of the particles was further confirmed by TEM (see Sect. 3.4.12 for details). The results show that the particle size of UCP is larger, and the average particle size is about 500 nm (see Fig. 3.4a for DLS and 3.4b for TEM); The particle size of UCN-S is the smallest, and its average particle size is about 20 nm (see Fig. 3.4c for DLS results and Fig. 3.4d for TEM results); The size between them is UCN nanoparticles. The average particle size is about 92 nm (see Fig. 3.4e for DLS results and Fig. 2.4 for TEM results). It can be seen that three nanomaterials with different nano sizes have different specific

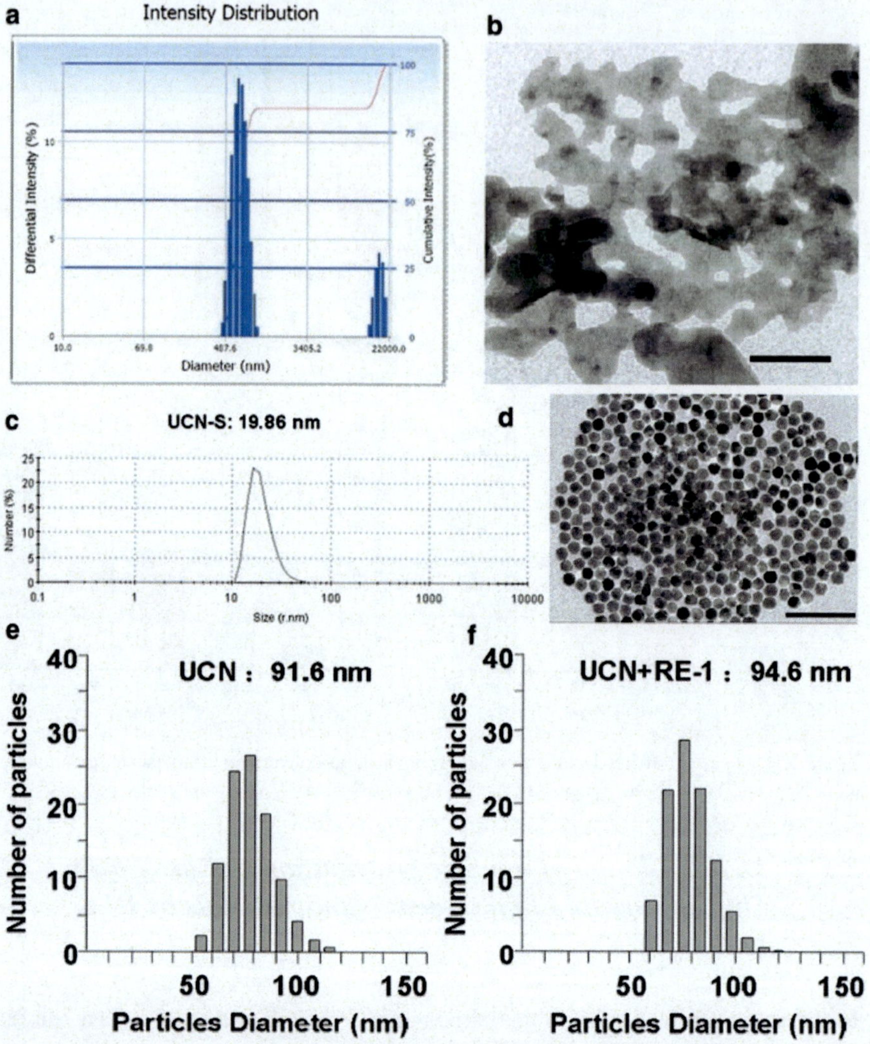

Fig. 3.4 Scale analysis of rare earth up-conversion luminescent nanomaterials

surface areas, which further explains the conclusion that UCN, UCP and UCN-S in Sect. 3.5.3 can combine different amounts RE-1 peptide.

Figure A is the particle size analysis of UCP, figure B is the TEM electron micrograph of UCP, figure C is the particle size analysis of UCN-S, figure D is the TEM electron micrograph of UCN-S, figure e is the particle size analysis of UCN, figure F is the particle size analysis of UCN coated with RE-1.

3.5.5 Calculate the Number of Binding Molecules of RE-1 and Rare Earth Up-Conversion Luminescent Nanomaterials

Nam-Hyuk Cho et al. [1] have reported that the average particle size of 100 μg iron oxide is about 15.7 nm. The amount of zinc oxide binding peptide that can be combined by zinc oxide core–shell nanoparticles is only about 1 μg (or 1 nmol). In contrast, the binding ability of RE-1 short peptide to rare earth nanomaterials is much higher than that of zinc oxide binding peptide to iron oxide zinc oxide core shell nanoparticles, and RE-1 short peptide binds more and more intensively on the surface of rare earth nanomaterials. In order to confirm this conclusion, we further calculated the amount of FITC-Re-1 short peptide on unit area rare earth up-conversion luminescent nanomaterials (NaYF4: Yb, Er). Since UCN and UCN-S are synthesized chemically and have regular spherical morphology under transmission electron microscope, we assume that UCN and UCN-S are similar to spheres. According to the volume formula of spheres, $V = \frac{4}{3} \times \pi R^3$, R is the radius of nanocrystals (UCN is about 46 nm, UCN-S is about 10 nm), and the record of 16–0334 in JCPDS card database (the density of β-$NaTF_4$ is ρ=4.309 g/cm^3), the average weight of each nanocrystal is $W = (\frac{4}{3} \times \pi R^3) \times \rho$ ($W_{UCN} = 1.76 \times 10^{-15}$ g, $W_{UCN\text{-}S} = 1.8 \times 10^{-17}$ g). According to Avogadro's constant NA = 3.023 × 10^{23} / mol, the molecular number of 100 μg UCN nanocrystals is 5.69 × 10^{10}; the molecular number of 100 μg UCN-S nanocrystals is 5.54 × 10^{12}. The surface area of the sphere is known to be (S) = 4 πR^2, so it can be seen that the surface area of a single UCN nanocrystal is 2.66 × 10^4 nm^2; the surface area of a single UCN-S nanocrystal is known to be 2.66 × 10^4 nm^2,its surface area is 1.256 × 10^3 nm^2. Therefore, the total surface area of 100 μg UCN nanocrystals is 1.51 × 10^{15}nm^2,100 μg UCN-S nanocrystals is 6.96 × 10^{15} nm^2. In experiment 3.5.3, we know that 100 μg UCN can bind to FITC-Re-1 47 μg (about 30 nmol), and 100 μg UCN-S can bind to FITC- Re-1 301 μg (about 192 nmol). The results show that a single UCN nanoparticle can bind 3.2 × 10^5 FITC-Re-1 peptide molecules, and a single UCN-S nanoparticle can bind 2.1 × 10^4 FITC- Re-1 peptide molecules.It is known that the molecular weight of fluorescence peptide FITC-RE-1 is 1553.8 Da, so the maximum number of FITC-Re-1 peptides that can be combined by the unit square nanometer UCN nanoparticles is 12.03, while the maximum number of FITC- Re-1 peptides that can be combined by the unit square nanometer UCN-S nanoparticles is 16.77.

3.5.6 The Effect of Control Peptide AP-1 on the Combination of RE-1 and Rare Earth Up-Conversion Luminescent Nanomaterials

In order to further confirm the high affinity binding ability of RE-1 short peptide and rare earth up-conversion luminescent nanomaterials, we designed a series of

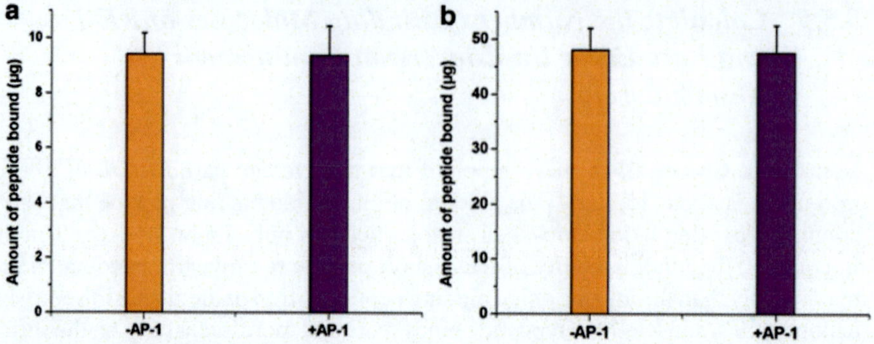

Fig. 3.5 Effect of AP-1 on the binding capacity of RE-1 and rare earth up-conversion luminescent nanomaterials. **a** is UCN, **b** is UCP

validation experiments. First of all, we tested the binding ability of RE-1 short peptide with UCN and UCP in the presence of control peptide by using the competitive method of control peptide (see Sect. 3.4.6 for details).To achieve the experimental results, we increased the concentration of AP-1 to 1000 times than RE-1 peptide, making it fully competitive with RE-1. The results show that AP-1 has no effect on the combination of RE-1 and rare earth up-conversion luminescent nanomaterials (as shown in Fig. 3.5). Furthermore, the high affinity binding ability of RE-1 with rare earth up-conversion luminescent nanomaterials was demonstrated.

3.5.7 *ζ Potential Analysis*

The above series of experiments have preliminarily proved the high affinity of RE-1 short peptide to rare-earth up-conversion luminescent nanomaterials, so how does RE-1 short peptide interact with UCN? It is known that RE-1 is a short peptide with 11 amino acids, and its isoelectric point is determined, namely $PI = 8.24$, which is positively charged in aqueous solution. Through the detection of UCN 5 potential, as shown in Fig. 3.6, the zeta potential of UCN in aqueous solution is $+22.6$, which is also positively charged. Interestingly, we found that the binding zeta potential of RE-1 combined with UCN was $+37.9$, with a higher positive charge than uncoated UCN. Under the binding conditions, both the RE-1 short peptide and UCN nanoparticles are positively charged particles, indicating that RE-1 binding to UCN is not through electrostatic interaction.

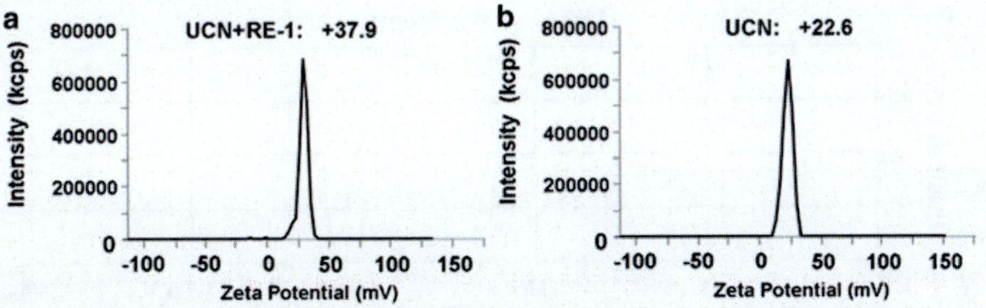

Fig. 3.6 ζ potential analysis. **a** UCN, **b** UCN wrapped with RE-1

3.5.8 Effect of RE-5 on UCN Particle Size and Potential

Experiment 3.5.7 results showed that the specific binding peptide RE-1 significantly increased the electric potential after encapsulating UCN, which was speculated to be due to the specific interaction between re-1 and UCN. To test this idea, we designed a controlled set of experiments. The effect of RE-5 on UCN particle size and 5 potential was verified by RE-5, a peptide similar to RE-1. The 200 ug re-5 was mixed with UCN and incubated with the same peptide binding method (see method 3.4.4 for details). The particle size and potential of the RE-5 and UCN binding products were detected using method 3.4.10. The results showed that RE-5 hardly changed the nanoscale of UCN (Fig. 3.7) and ζ potential (Fig. 3.8), indicating that RE-5 could not bind UCN, which was consistent with the comparison of the binding ability of peptide analogs and nanometer Nd_20_3 obtained in the previous chapter (see Table 2.1).

3.5.9 Peptide Competition Proves the High Affinity of RE-1 Binding to UCN

RE-5, an analogue of RE-1 with only one amino acid change, was introduced as the control. In order to facilitate the detection, we designed a peptide competition experiment using fluorescent peptide (FITC-RE-1) and non-fluorescent peptide (RE-5 andRE-1) (see method in Sect. 3.4.11) to compare the binding ability of RE-1 and RE-5 to UCN respectively. The results in Fig. 3.9 show that even the excessive RE-5 short peptide has a very weak effect on FITC-RE-1 binding ability with UCN, while the RE-1 short peptide under the same conditions almost completely inhibits FITC-RE-1 binding ability with UCN. It was further proved that the combination of RE-1 and UCN has high affinity and specificity.

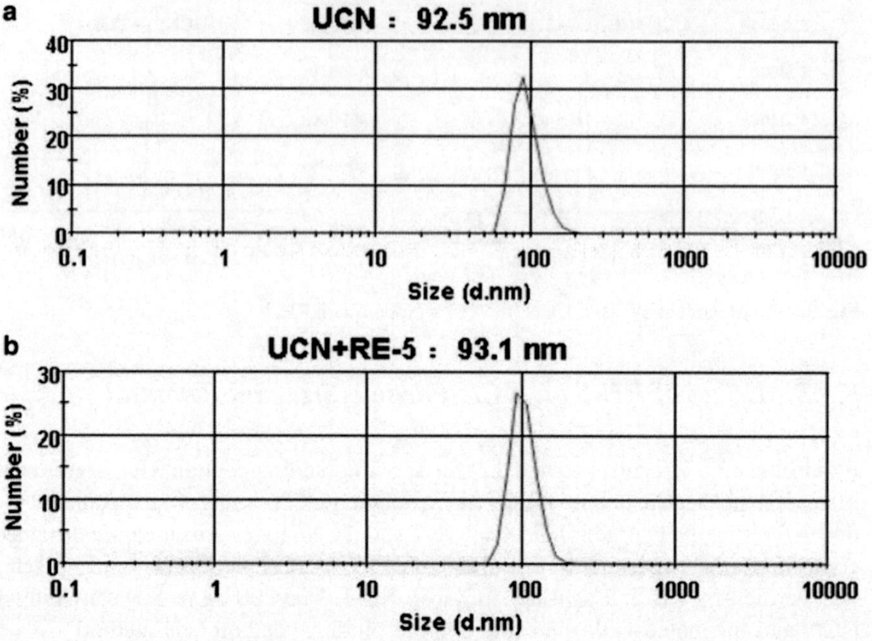

Fig. 3.7 DLS particle size analysis. **a** UCN, **b** UCN + RE-5

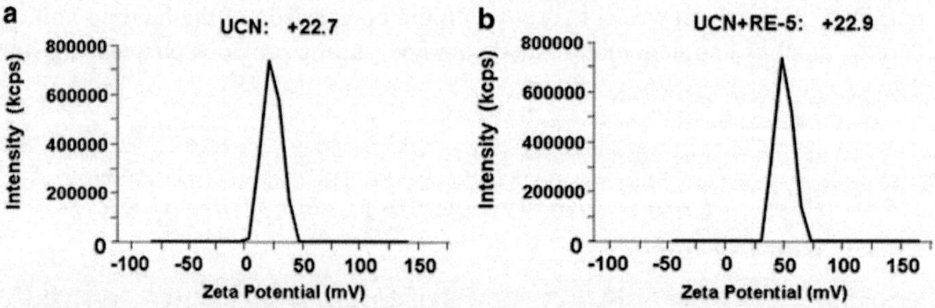

Fig. 3.8 ζ potential analysis. **a** UCN, **b** UCN wrapped with RE-5

3.5.10 Influence of Solution Environment on the Binding Ability of RE-1 to Rare Earth Up-Conversion Luminescent Nanomaterials

By comparing the binding ability of RE-1 short peptide and rare earth up-conversion luminescent nanomaterials (UCN, UCP) in different solution environments (PH, salt ion concentration, temperature), the high affinity binding ability of RE-1 and light-conversion luminescent nanomaterials on dilute soil was further determined.

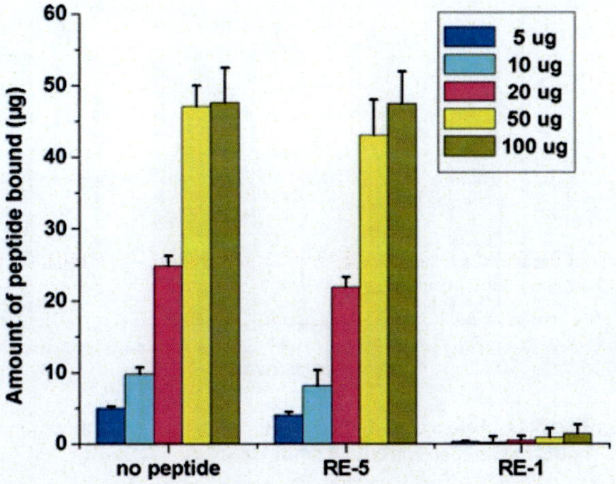

Fig. 3.9 Competition between fluorescent peptides and non-fluorescent peptides

First, the effects of different PH values on the binding capacity were tested. The results showed that the combination of re-1 with rare-earth up-conversion luminescent nanomaterials (UCN and UCP) was not sensitive to PH. We found that the amount of FITC-RE-1 bound to the surface of the luminescent nanomaterials (UCN, UCP) was almost the same on the rare earth conversion under acidic conditions (PH 3 and 5), neutral conditions (PH 7), and alkaline conditions (PH 10) (Fig. 3.10). Subsequently, we also tested the combination of RE-1 with rare earth up-conversion luminescent nanomaterials (UCN, UCP) in different salt ion concentrations (0, 0.15, 0.3, 0.5 M) (as shown in Fig. 3.11) and at different temperatures (25, 37, 55, 80 °C) (as shown in Fig. 3.12). This indicates that the combination of RE-1 with rare earth up-conversion luminescent nanomaterials (UCN, UCP) is very tight, this phase nanomaterials with high affinity and forms surface coating interactions that do not change with the surrounding solution environment.

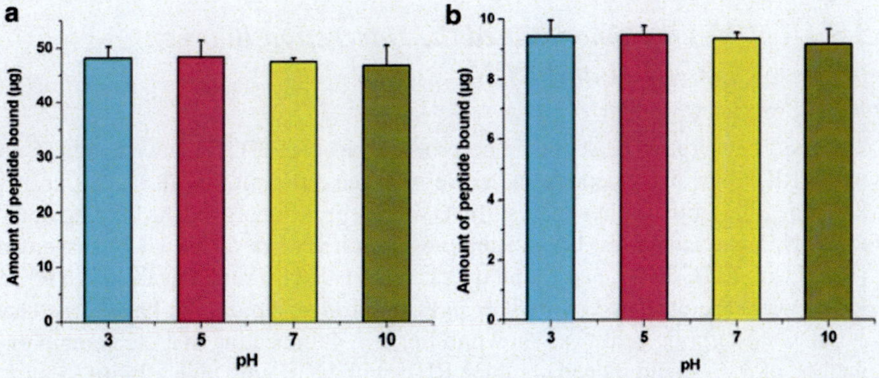

Fig. 3.10 Effect of PH on the combination of RE-1 and rare earth up-conversion optical nano materials. **a** UCN, **b** UCP

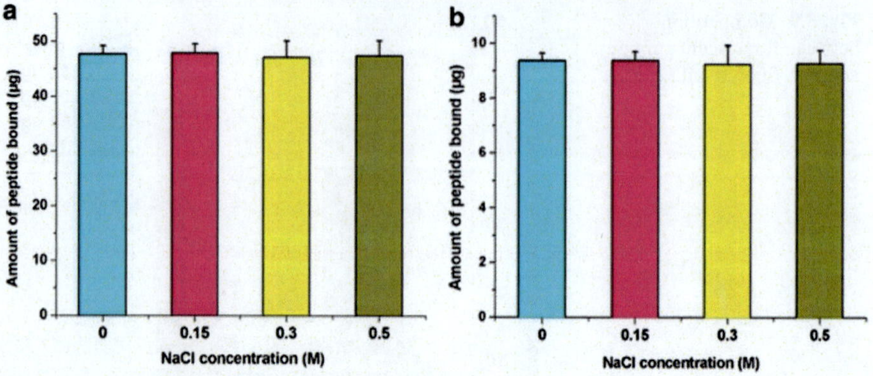

Fig. 3.11 Effect of salt ion concentration on the combination of RE-1 and rare earth up-conversion optical nano materials. **a** UCN, **b** UCP

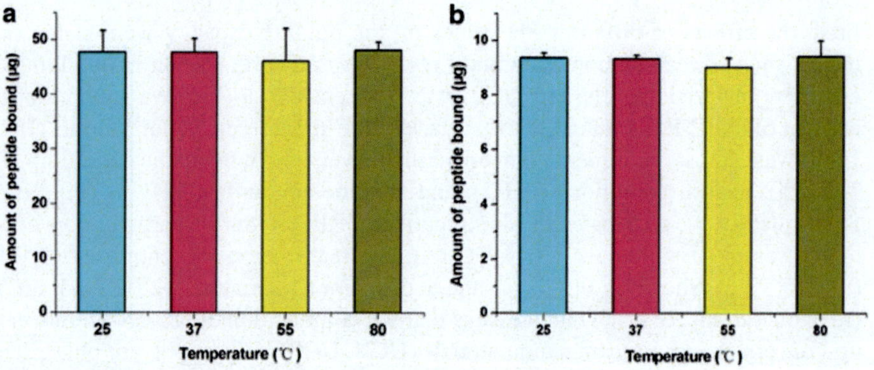

Fig. 3.12 Effect of temperature on the combination of RE-1 with rare earth up-conversion luminescent nanomaterials. **a** UCN, **b** UCP

3.5.11 UV–Vis Demonstrated the Interaction Between FITC-RE-1 and UCN

The visible absorption peak of the fluorescent molecule FITC is known to be 485 nm, while RE-1 is a small peptide molecule with the absorption peak of 230 nm and 280 mm. Therefore, further analysis by UV–Vis spectrum proved the binding of re-1 to UCN. We scanned the UV–Visible absorption spectra of the UCN suspension coated with FITC-RE-1, and found the characteristic absorption peaks of FTIC and RE-1, while the uncoated pure UCN in the control group did not have these characteristic absorption peaks (as shown in Fig. 3.13), indicating that the combination method of 3.4.4 could indeed combine RE-1 with UCN with high affinity.

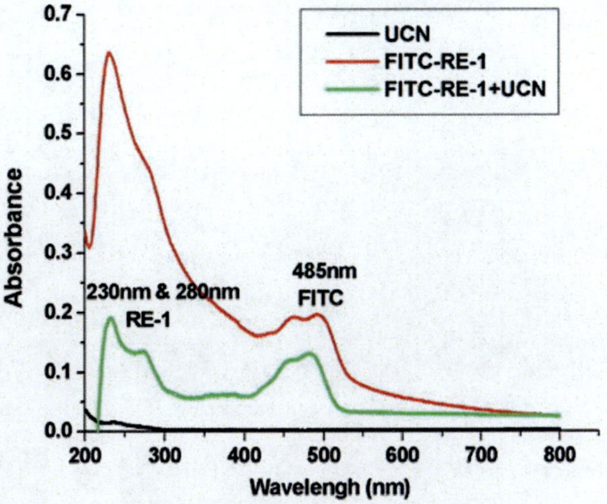

Fig. 3.13 UV–Vis absorption spectrum

3.5.12 *RE-1 Forms a Tight Peptide Coating on the Surface of UCN*

The above series of experiments have verified that RE-1 short peptide can specifically bind rare earth to convert luminescent nanomaterials, and this binding has a very high affinity. Next, we will further explore its binding form and junction mechanism. In the previous chapter, we used transmission electron microscopy (TEM) to observe the binding form of reob-1 phagocytes to UCN. However, due to the fact that phage REOB-1 is a virus particle, while REOB-1 is only a small polypeptide molecule composed of 11 amino acids, it cannot be concluded that the binding form of phage REOB-1 and UCN is the binding form of RE-1 short peptide and UCN based on their different properties. Through negative staining of RE-1 short peptide at different concentrations, the binding form of RE-1 to UCN was directly observed by transmission electron microscopy (TEM). Interesting, we found that a large number of RE-1 short peptide absorption winding in UCN surface, and in almost like a round ball UCN form a peptide parcel layer on the surface of nanoparticles (as shown in Fig. 3.14), in order to verify the results, and then we had a scanning electron microscope (SEM) observation, can be seen clearly under the scanning electron microscopy (SEM), not package UCN nanoparticles surface is smooth, and wrapped the RE-1 UCN rough surface is rough, and there are many filamentous crack, also confirmed the presence of the peptide UCN surface RE-1 layer (Fig. 3.15)Based on the analysis of the particle size of nanoparticles on the two types of electron microscopy, the particle size of UCN was about 90 nm, while the particle size of UCN coated with RE-1 was slightly larger, further proving the existence of the surface peptide layer, which was consistent with the previous results of DLS (see Fig. 3.4e for details).Therefore, it can be concluded that the RE-1 short peptide forms a tight peptide envelope on the surface of UCN.

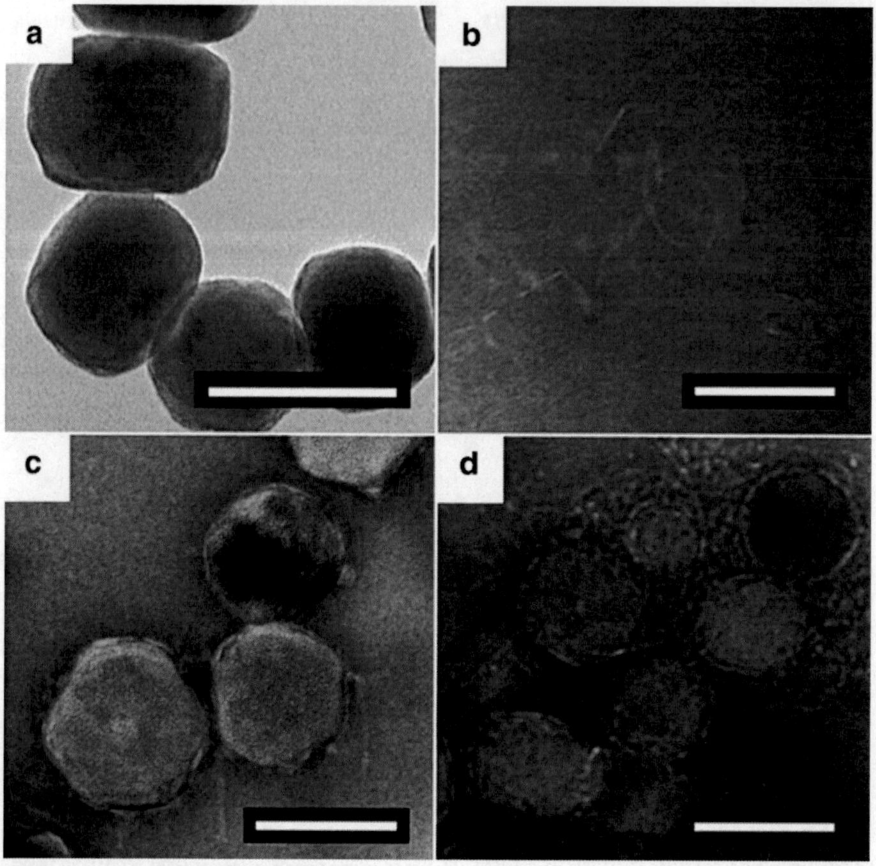

Fig. 3.14 TEM image of UCN combined with RE-1. **a** UCN, **b** RE-1, **c** UCN wrapped with low concentration RE-1, **d** UCN wrapped with high concentration RE-1

3.5.13 Nuclear Magnetic Resonance Method (^{1}H NMR) Proved the Interaction Between RE-1 and UCN

Through the above electron microscopy, we found that the specific binding short peptide RE-1 could form a tight peptide envelope on the surface of UCN. To further confirm this conclusion, we conducted a series of verification experiments on the interaction between RE-1 short peptide and UCN. Nuclei, like electrons, have spin angular momentum. In the applied magnetic field, the interaction of nuclear spin angular momentum will change the nuclear spin precession of energy distribution among the magnetic field, can cause level splitting, causing the NMR spectra of signal bee shape changes, through analyzing the change of the peak shape, can infer a connection to the relationship between molecular structure of each atom. For isolated nuclei (H), the same kind of nuclei in the same strength of the external magnetic field,

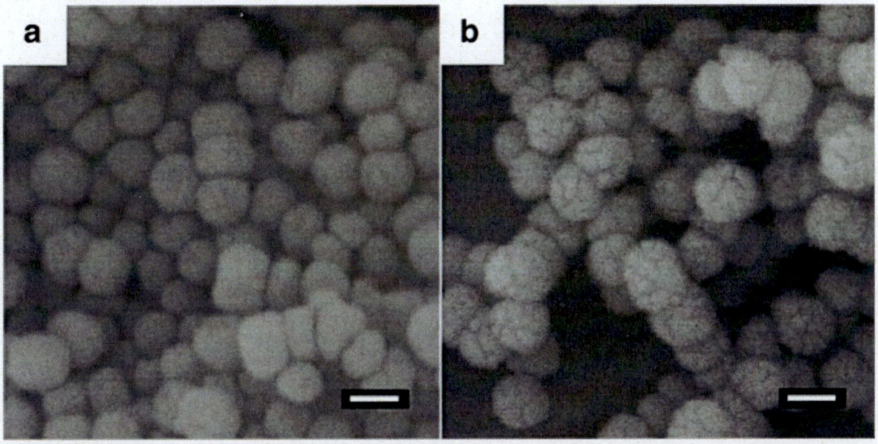

Fig. 3.15 SEM image of UCN combined with RE-1. **a** UCN, **b** UCN wrapped with RE-1

only is sensitive to a particular frequency of field but in the nucleus of the molecular structure, due to the influence of such factors as the electron cloud distribution in the molecule, real feelings to tend to a certain degree of changes in the external magnetic field strength, and in the molecular structure of different position of the original nucleus, and he felt the strength of the magnetic field are different, different signal peak appeared. However, the nuclei in the same chemical environment will show the same signal peak in NMR spectrum, and the NMR signal frequency caused by the influence of chemical environment will produce chemical shift. Therefore, ^{1}H NMR is often used to analyze the changes of substructure and group environment after molecular interaction. Short peptide in this experiment, RE-1 ^{1}H NMR signal peak in the same chemical environment is fixed, when combined with UCN RE-1, RE-1 in the changes of the chemical environment around, can lead to RE-1 ^{1}H NMR signal peak to produce chemical shift, so we can detect combines UCN RE-1 ^{1}H NMR chemical shift of signal peak to assess and RE-1 UCN interaction. We added 10ug/mL of RE-1 and 100 ug/mL of UCN coated with 10 μg/mL of RE-1 to 10% D20 and used ^{1}H NMR spectrometer to scan NMR one-dimensional hydrogen spectrum at 298 K. Results as shown in Fig. 3.16, the spectra of the signal peak of RE-1 and RE-1 combined with UCN were almost identical, but the chemical shift of the signal peak to the lower field at about 8.0 ppm in the RE-1 combined with UCN occurred at 0.05 ppm (the change of chemical shift is shown in the blue box in the figure). This shows that the amino acid molecules in RE-1 short peptide interact with UCN, which changes the chemical environment such as the chemical bond and the distribution of electron cloud near the H nucleus of RE-1 amino acid. It can be proved that RE-1 is combined with UCN.

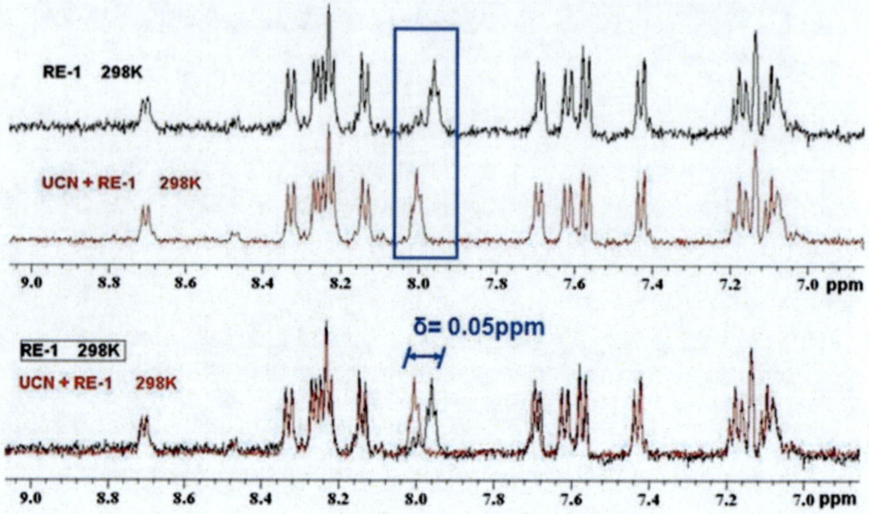

Fig. 3.16 Nnuclear magnetic resonance one-dimensional hydrogen spectra (^{1}HCNMR) of UCN combined with RE-1

3.5.14 Fourier Transform Infrared Spectroscopy (FTIR) Demonstrates the Interaction Between RE-1 and UCN

Then, the combination of RE-1 and UCN was further verified by Fourier transform infrared spectroscopy (FTIR).Source frequency of continuous change into the interference light, infrared light through the interferometer by irradiation with a group of samples, the samples of the molecules absorb some of the frequency of radiation, the molecular vibration or rotation causes the net change of dipole moment, make the vibration energy level transition from the ground state to the excited state, the interference spectra by fast Fu Li leaf transformation, light transmittance and absorption strength with the frequency or wave number transform infrared spectrogram. According to the spectral absorption band of month in the location, intensity and shape, use groups to research the relationship between vibration frequency and molecular structure of fixed absorption band, confirm or key groups in the molecules, and then by its characteristic vibration frequency shift, band intensity and the change of the shape to infer molecular structure and molecular interactions. In the infrared spectrum, the wave number in the interval of 4000–1330 cm^{-1} (wavelength 2.5–7.5 μm) is called the characteristic frequency region, or the characteristic region for short. The absorption peak in the characteristic area is sparse and easy to identify. The characteristic frequency of functional groups in various compounds is located in this region, where the vibration frequency is relatively high and is less affected by the rest of the molecule, so there are obvious characteristics, which can be used as the main basis for the characterization of functional groups. Since RE-1 is a small peptide molecule and contains a large number of groups with characteristic absorption band, we can determine the interaction between UCN and RE-1 according to the changes of RE-1 infrared band after binding to UCN. We performed infrared spectroscopy for UCN, which was uncoated and coated with RE-1 short peptide, and re-1 short peptide, respectively. After further analysis of the characteristic absorption peaks in the spectrum, it can be seen that the three samples all have a very wide and strong peak at the wave number of 3400 cm^{-1}, which is because the solvents of the three samples are all water stretching oscillation peak; There are almost no chemical groups on the surface of UCN; RE-1 is composed of amino acids, so the wave number at 1561 cm^{-1} is the characteristic absorption peak of C=O in the amino acids of RE-1 short peptide. A small peak at 2569 cm^{-1} may be the S–H stretch vibration peak of cysteine in the 11 amino acids of RE-1.However, the s–h stretch vibration peak at 2569 cm^{-1} disappeared in the sample after RE-1 was combined with UCN, indicating that the combination of UCN and RE-1 did occur. At the same time, by comparing the spectra before and after the combination of RE-1 and UCN, it was found that the wave number shifted by 40 cm between the two strong peaks of 2100 and 2140 cm^{-1}, which further indicated that the combination between UCN and RE-1 took place (as shown in Fig. 3.17).

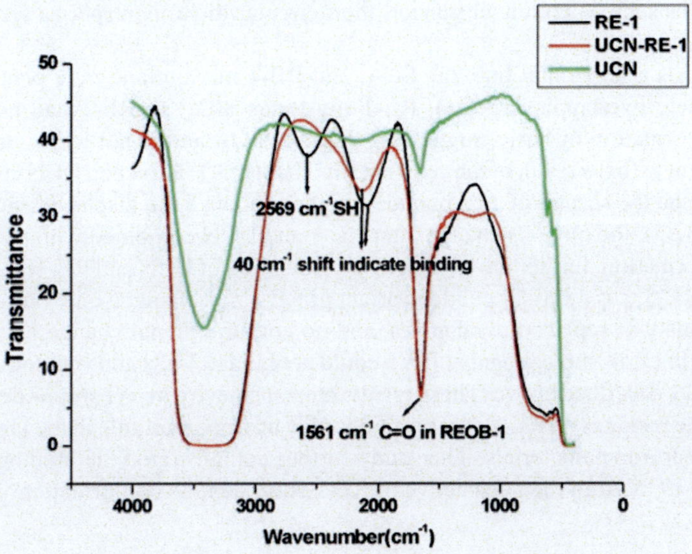

Fig. 3.17 Fourier transform infrared spectra of UCN combined with RE-1

3.5.15 Circular Dichroism Spectroscopy (CD) Proves the Interaction Between RE-1 and UCN

Circular dichroism has unique advantages in determining the configuration and conformation of chiral compounds and determining the position of functional groups in chiral molecules. Chiral material (optical activity of molecules) absorption degree of left and right circularly polarized light, the amplitude of the output when the electric field vector by different samples, the synthesis is elliptically polarized light, polarized light again according to record the spectrum of chiral molecules found base group's configuration and conformation, it is widely used in analytical research of interaction between protein and protein. In this experiment, we can use circular dichroism to study the specific interaction between UCN and RE-1 from the conformation of RE-1.We will combine with UCN the RE-1 and not pure peptide RE-1 scanned for round two color spectrum, respectively, at the same time set in the control group: not UCN, using the same method of package against the UCN the peptide RE-5, according to the peptide RE-5, SiO_3 contrast material, and using the same method package RE-1 SiO_3, under the condition of same scanned for round two color spectrum, respectively. According to the scanning results, because nanomaterials UCN and SiO_3 are inorganic nanomaterials and do not belong to optically active molecules such as chiral substances, their circular dichroism spectra are similar to the background.

Basic is a horizontal line: the SiO_3, and RE-1 mixture and pure peptide almost completely overlapping spectra, RE-1 illustrates SiO_3) to RE-1 had no effect on the conformation of basic, suggesting that the SiO_3, and is not RE-1, this result is consistent with the result of the center of this chapter 3.1: RE-1 and UCN combination spectra and the spectra of pure peptide happened about 5 nm displacement (as shown in Fig. 3.18), the obvious change after RE-1 and UCN combined with the change in its conformation, further proves that the combination of UCN and RE-1;In the control group, RE-5 was a similar peptide variant of RE-1 that changed only one amino acid (threonine was replaced by alanine), and no conformational change occurred after binding to UCN, indicating that RE-5 could not bind to UCN and was consistent with the results described above. These results indicate that hydroxyl groups derived from threonine residues on RE-1 short peptides play an important role in the identification of rare earth nanomaterials. This study further demonstrated the binding ability of RE-1 to UCN from the perspective of RE-1 short peptide conformation.

3.5.16 The Interaction Between RE-1 and UCN Was Demonstrated by Isothermal Drop Quantitative Thermal Method (ITC)

Isothermal drop calorimetry is an important method for the study of bio-thermodynamics and biodynamics developed in recent years. It can obtain the

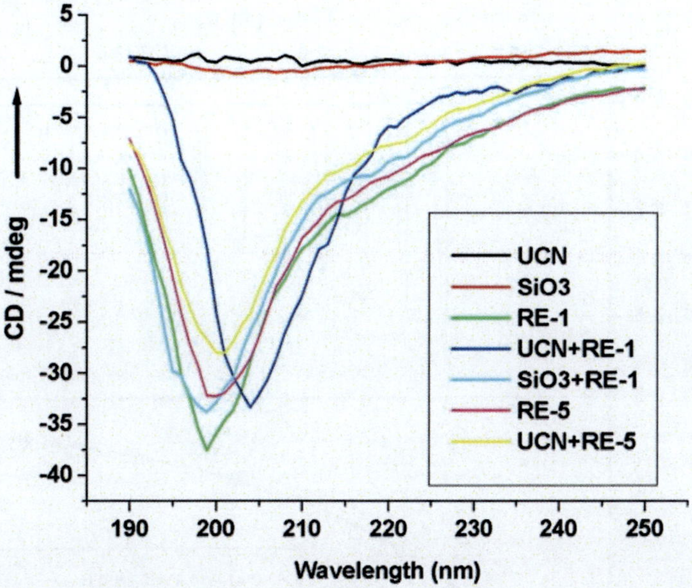

Fig. 3.18 Circular dichroism spectra of UCN combined with RE-1

complete thermodynamic parameters of biomolecular interactions by continuously and accurately monitoring and recording the calorimetric curves of each change process with a highly sensitive and automated microcalorimeter. We can evaluate the interaction between molecules based on the obtained thermodynamic parameters, further calculate the thermodynamic constant of the combination of UCN and RE-1, and evaluate the interaction force between RE-1 and UCN. Through the ITC titration experiment, we studied the combination of UCN and RE-1, as shown in Fig. 3.19. The upper part of the curve is the change of the heat of the ITC titration with time, and the lower part is the change of enthalpy obtained by integrating the titration curve with the titration ratio. After analyzing the data, it can be concluded that UCN is bound to RE-1 at 1:8.46, and the apparent affinity KD, ΔH and ΔS are 4.31E4 $\pm$ 1.30E4M^{-1}, -2.403 $\pm$ 0.1663 kcal/mol, 5.39 $\pm$ 0.48 kcal/mol, 13.1 cal/(mol*K). It is further proved that RE-1 has a high affinity for UCN.

3.5.17 Surface Plasmon Resonance (SPR) Method Proved the Interaction Between RE-1 and UCN

SPR is a new technology that uses the principle of plasmon resonance to detect the interaction between ligands and analytes on biosensor chips. In recent years, SPR has been widely used to detect the binding effect between biomolecules, or to complete

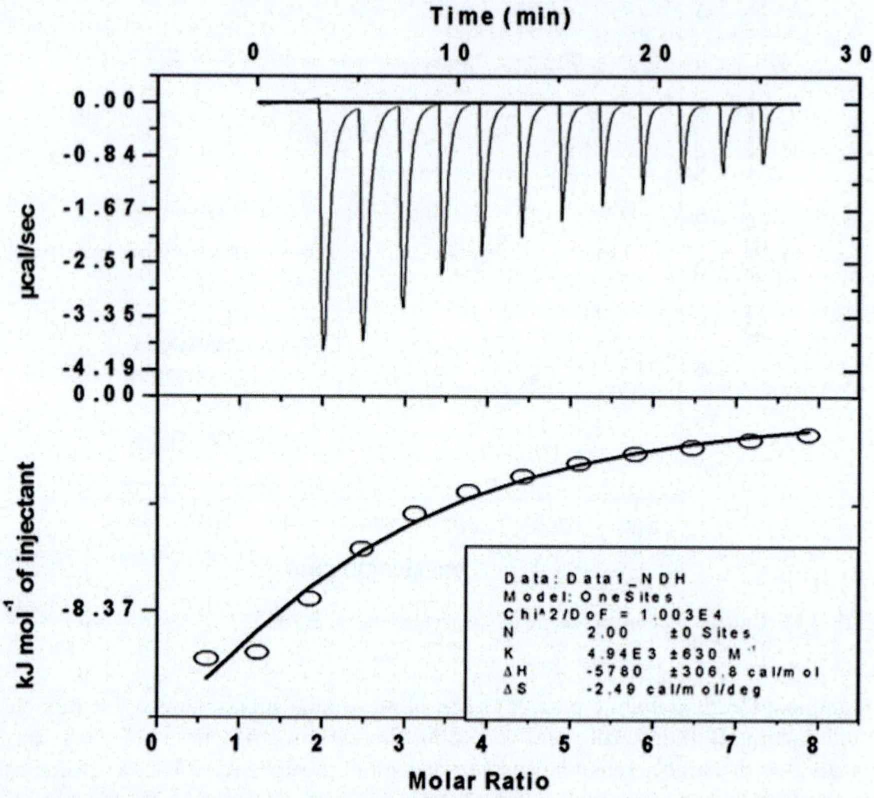

Fig. 3.19 Thermodynamic constants of UCN binding to RE-1 were calculated by ITC titration

the identification of specific biomolecules and the determination of their concentration through the detection of biomolecular junction cooperation. We also used SPR to study the combination of RE-1 and UCN. We coupled the RE-1 short peptide to the surface of the sensor chip, and then added the suspension of UCN nanoparticles to make RE-1 capture UCN. The mobile phase of the control group replaced UCN with the solvent milli-Q water as basicline. As shown in Fig. 3.20, the blue curve is the binding curve of UCN and RE-1 after deducting the control group, and the black box shows the strong junction peak after deducting the control group, indicating that RE-1 can capture the UCN in the mobile phase, further proving that UCN has a strong binding force with RE-1.It is worth mentioning that only preliminary experiments in SPR detection are described here. Most biomolecular interactions exist in the equilibrium of binding and dissociation, and SPR can simultaneously detect the binding and dissociation constants of intermolecular interactions. In this experiment, the short peptide RE-1 was used as the stationary phase, and UCN was used as the flowing phase for capture. When the chip was rinsed with milli-Q water, a solvent that diluted RE-1 and UCN, RE-1 and during the dissociation of UCN, we found that it was impossible to elute UCN from the chip coupled with RE-1 with milli-Q

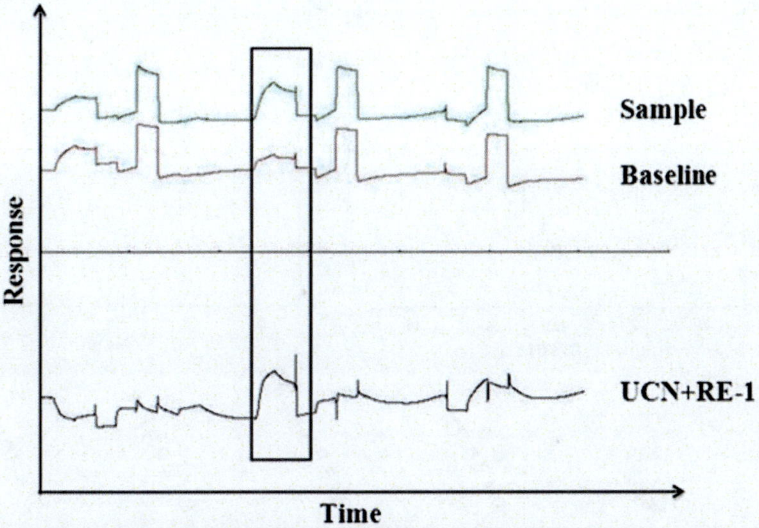

Fig. 3.20 SPR verifies the combined effect of UCN and RE-1

water, and it was still impossible to dissociate UCN from RE-1 after trying eluents such as weak acid and weak base, so the dissociation constant could not be obtained. This further demonstrates the extremely high binding ability of RE-1 to UCN.

3.5.18 Analysis of Dissociation Conditions After the Combination of RE-1 and Rare Earth Up-Conversion Luminescent Nanomaterials

The above series of validation experiments have proved that specific binding short peptide RE-1 can transform luminescent nanomaterials on thin soil with high affinity. In the SPR test of 3.16, due to the high binding force of RE-1 and UCN, UCN cannot be dissociated from the chip coupled with RE-1. To further confirm this result we tried the following dissociation experiment. We first examined the dissociation of RE-1 combined with UCN under static conditions. To facilitate the detection, FITC-RE-1 combined with UCN was placed at room temperature and centrifuged at different time points to detect the amount of FITC-RE-1 dissociated fluorescent peptide in the supernatant. As shown in Fig. 3.21a, RE-1 is highly stable bound to UCN and dissolves less than 0.7% after 48 h of standing. Similarly, RE-1 is also stably bound to UCP, and the dissociation amount of RE-1 is less than 1.5% after 48 h of standing (as shown in Fig. 3.21b). Subsequently, we further tested the effect of different severe treatment methods (detergent SDS, boiling, strong acid, etc.) on the binding ability of RE-1 to UCP. The results are shown in Fig. 3.21c. Among the treatments, RE-1 and

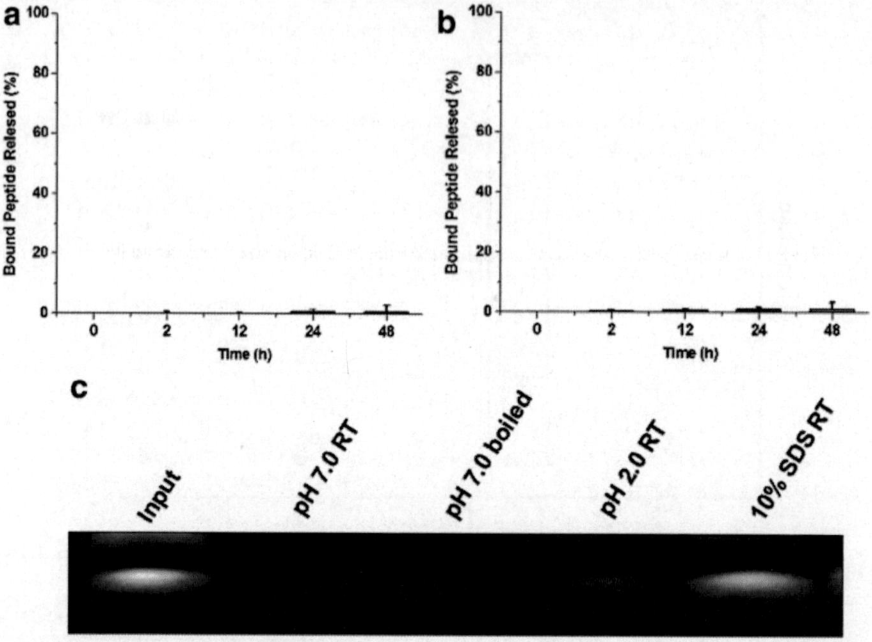

Fig. 3.21 Detection of dissociation between FITC-RE-1 and rare earth converted luminescent nanomaterials. **a** The dissociation ratio of FITC-RE-1 in UCN when FITC-RE-1 is wrapped after standing at different times. **b** The dissociation ratio of FITC-RE-1 in UCP when FITC-RE-1 is wrapped after standing at different times. **c** The results of electrophoresis, showing the dissociation of FITC-RE-1 wrapped UCN after treatment under different conditions. From left to right, the treatment sequence was: FITC-RE-1 control group, 50 mM Tris–HCL (pH 7.0) group at room temperature, and 50 mM Tris–HCL (pH 7.0) group

UCP can be dissociated only after the treatment with anion detergent SDS. However, neither strong acid nor boiling can break the binding of RE-1 to UCP, which further indicates the high affinity binding of RE-1 to rare earth up-conversion luminescent nanomaterials.

Boiling treatment group, glycine–HCL (pH 2.0) treatment group at room temperature, 10% SDS treatment group.

3.6 Summary of this Chapter

In this chapter, a series of validation experiments were used to demonstrate the specific and highly affinity binding of RE-1 short peptide to rare earth nanomaterials. RE-1 short peptide has material specificity. It is a specific binding peptide for broad-spectrum rare earth nanomaterials, while it is not binding for non-rare earth nanomaterials. Through electron microscopic observation, we found interestingly that RE-1 short peptide could bind to form a tight polypeptide coating on the surface of rare earth nanomaterials. Rare earth up-conversion luminescence nanomaterials have unique up-conversion luminescence characteristics and great application potential in the field of biomedicine, which makes them stand out among many nanomaterials. In order to better study and expand the application of RE-1 polypeptide in rare earth nanomaterials, UCN, UCP and UCN-S are mainly selected in this paper. Through the analysis of the particle sizes of these three materials, we found that UCP particles are relatively large, with an average particle size of about 500 mm. UCN-S particles were the smallest, with an average particle size of about 20 nm. In between are UCN nanoparticles with an average particle size of about 92 nm. Because of UCN and UCN-S For chemical synthesis, the morphology is relatively regular, so we calculated that the maximum number of FITC-RE-1 peptide molecules that can bind to UCN nanoparticles per unit square nanometer is 12.03, while the maximum number of FITC-RE-1 peptide molecules that can bind to UCN-S nanometers per unit square nanometer is 16.77.Through a series of verification experiments such as ^{1}H NMR, FTIR, SPR, CD and ITC, we further proved that the combination of RE-1 with rare earth up-conversion luminescent nanomaterials is very tight. Through the strength dissociation experiment, we found that neither strong acid nor boiling could break the combination of RE-1 with UCP, and the interaction between RE-1 and rare earth nanomaterials did not change with the surrounding solution environment. It is further demonstrated that RE-1 is highly affinity to RE-1 converted luminescent nanomaterials.

Reference

1. Cho, N. H., et al. (2011). A multifunctional core-shell nanoparticle for dendritic cell-based cancer immunotherapy. *Nature Nanotechnology, 6*(10), 675–682.

Chapter 4
RE-1 Improves the Suspension Capacity of Rare Earth Nanomaterials in Water Well Reduction and Cell and Surface Interactions

4.1 Introduction

Rare earth up-conversion luminescent nanomaterials (UCN, UCP, UCN-S) emit light due to their unique up-conversion luminescence characteristics, to make a wide range of applications, after the field of biomedical chase. In order to further enhance the application of the materials themselves in various fields (especially in the field of biomedicine), the short peptides of specific conjugated thin soils for the conversion of luminescent nanomaterials were screened in this paper. Therefore, whether the inclusion of RE-1 has an impact on the optical properties of rare-earth up-conversion luminescent nanomaterials deserves our special attention. In previous work, we found that on rare earth conversion luminescent nano materials (such as UCN, UCP contains, UCN-S) adhesion ability is very strong, the experiment used in the test bench, centrifugal tube multi-wave publish or perish top surface of the medium, after the 980 nm laser scanning, can show green light, the visible environment to all adhere to these materials, especially containing rare earth conversion luminescent nanomaterials of centrifugal pipe wall adhesion of materials; However, none green fluorescent material was found on the wall of the centrifuge tube containing the rare-earth up-conversion luminescent nanomaterials coated with RE-1. We also detected the nozzle head that had absorbed the above combination products, and no residue of green fluorescent material was found. Therefore, in this chapter, we will describe the effect of RE-1 on the non-specific adhesion of rare earth up-converted luminescent nanomaterials by using the optical properties of rare earth up-converted luminescent nanomaterials, and further identify the changes in the physical and chemical properties (such as suspension properties) of rare-earth up-converted luminescent nanomaterials coated with RE-1.

Y. Zhang, *Tuning Autophagy-Inducing Activity and Toxicity for Lanthanide Nanocrystals*, Springer Theses, https://doi.org/10.1007/978-981-16-8166-0_4

4.2 Materials

4.2.1 Cell Lines

HeLa: human cervical cancer cell line (purchased from Shanghai institute of biochemistry, Chinese academy of sciences).

4.2.2 Reagents

4% paraformaldehyde: 8.5 g NaCl and 30.8 g $NaHPO_4$, $12H_2O$, 2.8 g NaH_2PO_4 were first weighed. $2H_2O$ was dissolved in 800 mL milli-Q water, adjusted pH value to 7.4 and then fixed PBS to 100 ml.

NMR buffer: a buffer containing 20 mM Na_2HPO_4 and 100 mM NaCl (pH 7.0) was prepared, weighed 3.5814 g Na_2HPO_4, 2.922 g NaCl, and dissolved in 450 mL of milli-Q water. Adjust the pH value to 7.0, and add additional milli-Q water to the total volume of 500 mL for later use.

4.2.3 Equipment and Consumables

Fluorescent detection colorimetric (X72052, ALPHA, UK).

UV detection pool (X72053, ALPHA, UK).

The cell culture dish and cell culture plate related to cell experiment were purchased from Corning company.

4.3 Instruments and Equipment

Fluorescence spectrophotometer (RF-5301PC, shimazu, Japan).
980 nm infrared laser (MDL-980 nm, 1 W, Changchun industry optoelectronic technology co. LTD. China).
fluorescence microscope (Olympus LX71, Olympus, Japan)
Ultraviolet spectrophotometer (DU-640, Beckman, USA)
Travelling vortex mixer (MSI Minishaker, IKA, China)
Trace Centrifuge (Centrifuge 5415D, Eppendorf, Germany)
Camera (Nikon D9000, Nikon, Japan)
Magnetic resonance spectrometer (Avance 500.132 MHz, Bruker, Germany).

4.4 Methods

4.4.1 Up-Conversion Fluorescence Spectrum Detection Method for UCN

4.4.1.1 Equipment Construction

Due to the xenon lamp built into the fluorescence spectrophotometer, the spectral range of the excitation light source is only 220–750 nm, while the rare earth up-conversion luminescence nanomaterials (UCN, UCN-S, UCP) used in this paper are up-conversion luminescence, that is, 980 mm infrared excitation and 540 mm green light emission. Therefore, to detect rare-earth up-conversion luminescence nanomaterials, an infrared laser (980 nm) capable of exciting infrared lasers must be equipped with a fluorescence spectrophotometer. Therefore, an infrared laser with a 980 mm excitation light source was connected to the external interface of the sample pool of the fluorescence spectrophotometer, and the detection could be made after sealing and shading. Note: due to the high heat of the infrared laser, in order to avoid the loss of optical components of the fluorescence spectrophotometer instrument when direct radiation, the power of the infrared laser should be turned off in a timely manner.

According to method 3.4.4, UCN was fully combined with RE-1, and the RE-1 coated UCN nanoparticles were re-suspended with 1 mL milli-Q water and then added to the colorimetric vessel. The results were measured and recorded with a fluorescence spectrophotometer equipped with an infrared laser. When measuring UCN, 100 L UCN (1 mg/mL) was added to 900 μL mill-q water to make the final volume reach 1 mL and was added to the colouring dish for measurement.

4.4.2 Cell Culture

All cells were cultured in a cell incubator at 37 °C containing 5% CO_2 with modified Eagle's medium (DMEM medium) and 10% fetal bovine serum, 100 U/mL penicillin and 100 μg/mL streptolycin were added to the medium.

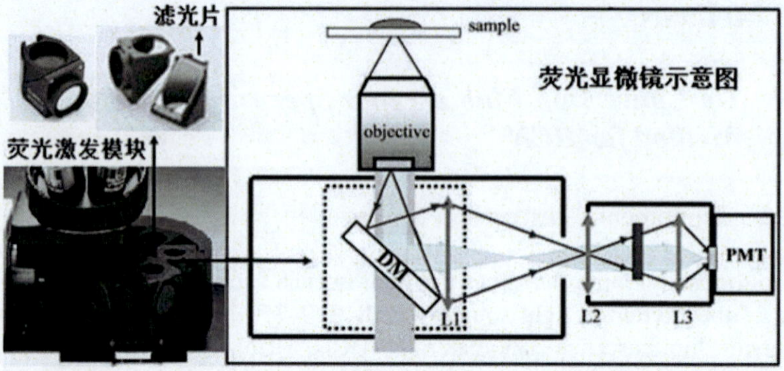

Fig. 4.1 Schematic diagram of fluorescence microscope

4.4.3 RE-1 Affects the Identification Method of Nonspecific Adhesion Ability of Rare Earth Up-Conversion Luminescent Nanomaterials

4.4.3.1 Equipment Construction

Because the Olympus LX71 fluorescence microscope only equipped with UV and blue light, green light, red light within four excitation light source, detect infrared excitation on rare earth conversion luminescent nanomaterials. Therefore, requires an external 980 nm infrared laser excitation light source. As a result of fluorescence microscopy imaging system has not been equipped with infrared filter, luminous infrared excitation light excitation sample photos after capture microscope imaging system, will present a dense infrared line on the sample grating. In order to solve this problem, we made with an infrared filter (to filter out more than 800 nm Fluorescent excitation module (as shown in Fig. 4.1), a special new module is added to the fluorescent excitation module to specially take photos of the converted luminescent nanomaterials on rare earth, which improves the signal-to-noise ratio of the images collected by the imaging system.

4.4.3.2 Test Methods

HeLa cells were inoculated into 24 holes in a 96-well cell culture plate with a density of 1–2 × 10, and cultured overnight in a cell incubator at 37 °C. The next day, 12 HeLa cells attached to the wall were transferred to fresh DMEM medium for later use. In addition, 12 well cells were washed twice with PBS and then fixed with 49% paraformaldehyde fixation solution for reserve. According to method 3.4.4, re-1 is wrapped on the surface of UCN (or UCN-S).With secrete degrees to 100 ug/mL without package and the RE-1 UCN (or UCN-S) processing 96 hole in the backup, fixed live HeLa cells and cell culture plate, processing after 2 h, PBS washing three times, then use equipped with 980 nm infrared laser fluorescence microscopy and HeLa living cells, cells and cell culture plate combined with fixed UCN (or UCN-S) nanoparticles.

4.4.4 Combination of PEG and UCN

The 10 μl PEG with different concentrations (high concentration 1 g/ml, low concentration 10 mg /ml) was added to 1 mg UCN (concentration 1 ng/ml), and the swirl oscillation was fully mixed. The samples were gently shaken and incubated at room temperature for 1 h and centrifuged at 12,000 pm for 10 min.

4.4.5 Methods for Time-Dynamic Detection of Settlement Rate

UCN of 3 mg was combined with 1.5 mg of RE-1 and its analogs, RE-2, RE-3 and RE-5, respectively according to the combination method of 3.4.4, and the excess uncombined RE-1 was washed off, and the precipitate was suspended in 3 mL water for backup. The 3 mg UCP was combined with 300 μg RE-1 and its analogs (RE-2, RE-3, RE-4, RE-5 etc.) and the control peptide AP-1 in accordance with the 3.4.4 binding method, the excess unbound RE-1 was washed off, and the precipitate was suspended in 3 mL water for later use. To 3 mg of UCN combined with PEG of 3 g in accordance with the method 4.4.4, the precipitation of nanoparticles with milli-Q water after washing twice, heavy suspended in 3 mL water, will be the preparation of samples (such as quantity and volume of package UCN, UCP contains, UCN-S as a control) moderate ultrasonic 1 min rear respectively in ultraviolet detection in the pool, with ultraviolet spectrophotometer at 500 nm absorption wavelength under continuous scanning 120 min, recording time dynamic curve.

4.4.6 Direct Observation of UCP Suspension Capacity

The 3 mL UCP (with a concentration of 1 mg/mL) was combined with 300 μg RE-1 and its analogs (RE-2, RE-3, RE-5 etc.) according to the 3.4.4 binding method, centrifuged at 12,000 rpm for 10 min, washed off the excess unbound RE-1 in the precipitate, and then re-suspended in 3 mL water. After 24 h of standing at room temperature, a hand-held 980 nm infrared laser was used to directly illuminate the upper and side of the colorimetric dish and take pictures with Nikon camera.

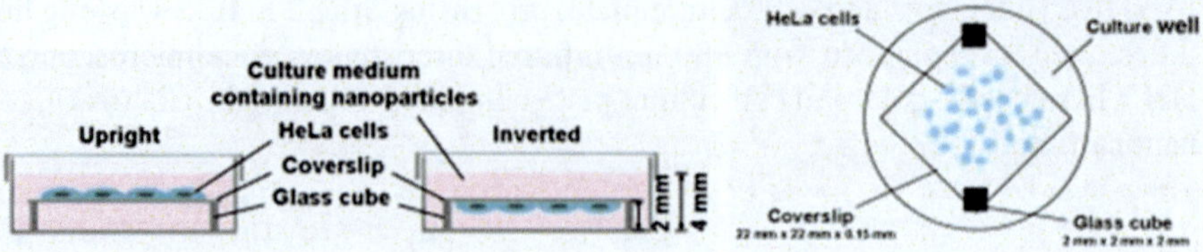

Fig. 4.2 Schematic diagram of diffusion ability detection

4.4.7 Test Method for Diffusion Capacity of Rare Earth Up-Conversion Luminescent Nanomaterials in Solution

4.4.7.1 Construction of Experimental Equipment

Cover glass slides (both 22 mm in length and width) and small cut vitreous bodies (both 2 mm in length, width and height) were put into a beaker, sterilized at 121 °C for 20 min, and dried in an oven for later use. In the inter-cell ultra-clean work table, two sterilized small vitreous bodies were burned with tweezers under the alcohol lamp, and then quickly placed on both ends of the round diameter of each hole of the six-hole plate (since the temperature of the small vitreous body was too high, it would just stick in the six-hole plate of plastic, and then it could be fixed) (as shown in diagram 4.2).

4.4.7.2 Test Methods

The sterilized cover glass was placed in a cell culture dish and dropped into HeLa cells to be cultured on the cover glass. When the cell fusion degree on the cover glass reached 95% or more, the cover glass with cells on one side facing up or down was placed in a 6-well plate and then mounted on a small vitreous body attached to the bottom of the 6-well plate. According to the method 3.4.4 RE-1 and their analogues RE-3, RE-3, RE-5 packages on rare earth conversion luminescent nano material surface, add 4 mL DMEM medium respectively to not package material and package already RE-1 in materials (the final concentration of rare earth on conversion luminescent nanomaterials of 100 μg/mL), mix after slow in six samples of orifice plate hole, and make full immersion medium on the cover glass, the location of the cell cover glass right in the middle of a liquid volume height (as shown in diagram 4.2).The cells were gently cultured in a 37 °C cell incubator for 2 h, and then the six-well plates were removed. Each sample hole was washed with PBS for 3 times, and the combined rare-earth up-conversion luminescent nanomaterials on the cell surface were observed with a fluorescence microscope equipped with a 980 mm infrared laser.

4.4.8 Data Analysis

All experiments were repeated at least three times. All data were represented by meant ± S.E.M., the student double-tailed t test. The data analysis was processed by two-tailed students test in the Origin data analysis software. The statistically significant difference was *$p < 0.05$, **$p < 0.01$, ***$p < 0.005$.

4.5 Results and Discussion

4.5.1 RE-1 Does not Affect Up-Conversion Fluorescence of UCN

In this paper, the short peptides of specific rare earth up-conversion luminescent nanomaterials were screened to further enhance the application of the materials themselves in various fields (especially in the field of biomedicine), so whether the inclusion of RE-1 has an impact on the optical properties of rare earth up-conversion luminescent nanomaterials is worth our attention. By detecting the up-conversion spectra of UCN coated with RE-1 and uncoated UCN, we found that the emission spectra and luminescence intensity of UCN coated with RE-1 were almost the same as that of pure UCN nanomaterials, with no effect at all (see Fig. 4.3).Based on this, we conducted the following series of experiments by using the optical properties of rare earth converted luminescent nanomaterials such as UCN.

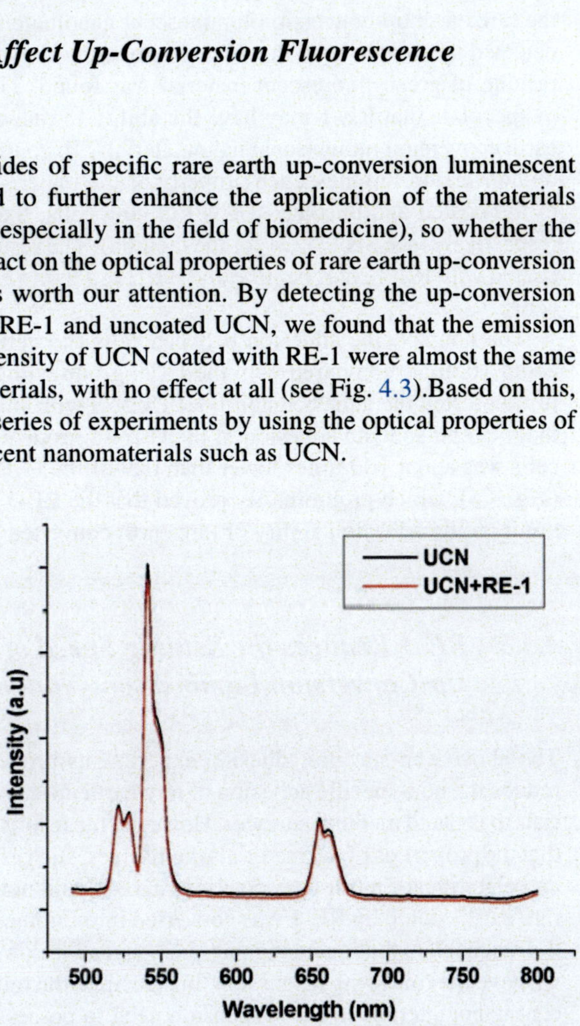

Fig. 4.3 Conversion fluorescence spectrum

4.5.2 RE-1 Reduces the Nonspecific Adhesion of Sparsely Converted Luminescent Nanomaterials

On the previous work, we found that the rare earth conversion luminescent nano materials (such as UCN, UCP, UCN-S) adhesive ability is very strong, the experiment used in the test bench, centrifugal tube, pipetting publish or perish the top surface of the medium, after the 980 nm laser scanning, can appear green, these materials filled the adhesion of the visible environment, especially containing rare earth conversion luminescent nanomaterials of centrifugal pipe wall adhesion of materials; However, no green fluorescent material was found on the wall of the centrifuge tube containing the rare earth up-conversion luminescent nanomaterials coated with RE-1. We also detected the nozzle head that had absorbed the above combination products, and no residue of green fluorescent material was found. This interesting finding leads us to speculate that RE-1 may have the ability to reduce nonspecific adhesion of rare earth conversion luminescent nanomaterials. To confirm this conjecture, we used the up-conversion luminescence property of the material to detect the adhesion ability of the coated and uncoated UCN to living cells, fixed cells and cell culture plates respectively (see Sect. 4.4.4 for the method). The results showed that the UCN group coated with RE-1 short peptide and the HeLa living cells were coated with fixed thin cells.

The non-specific adhesion between cells and cell culture plates was reduced by about 10 times compared with the UCN group without RE-1 coating (Fig. 4.4).We further tested the adhesion ability of UCN-S with small particles to cells, and found that the non-specific adhesion of the UCN-S group coated with RE-1 to Hela living cells was about 100 times lower than that of the UCN-S group uncoated with RE-1 (Fig. 4.5), which preliminarily proved that the RE-1 short peptide could reduce the non-specific adhesion ability of rare earth converted luminescent nanomaterials.

4.5.3 RE-1 Reduces the Settling Speed of the Rare Earth Up-Conversion Luminescent Nanomaterials

The above non-specific adhesion experiments proved that RE-1 short peptide can reduce the non-specific adhesion of rare earth up-conversion luminescent nanomaterials to cell and medium surfaces. However, the results in the previous chapter showed that the properties of the materials themselves, such as particle size, optical properties and electrification (all with positive charges), did not change significantly when the rare earth coated by RE-1 was converted into luminescent nanomaterials. So what's the mechanism that's causing this change in adhesion? In order to solve this problem, we have the following discussion. In principle, the rare earth up-conversion luminescent nanomaterials in the suspension need to possess two main abilities simultaneously: sedimentation and diffusion in order to contact cells growing at the bottom of 96-well cell culture plates in a solution environment. Based on this, we first detected

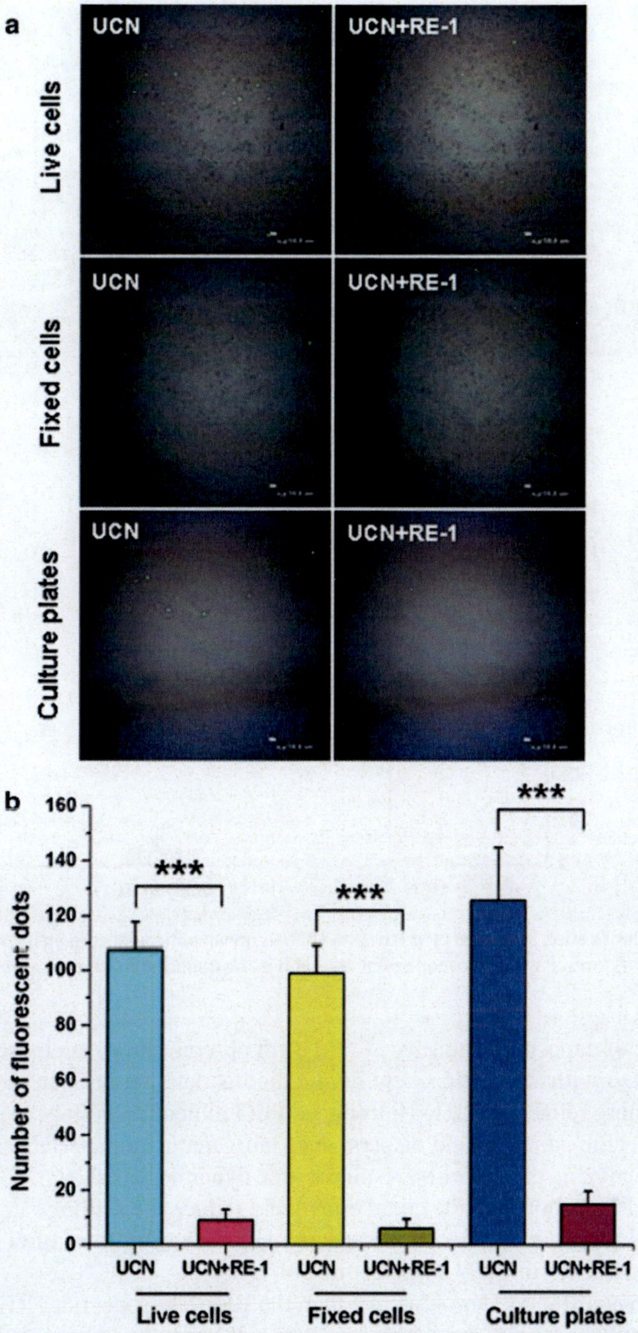

Fig. 4.4 Identification of the effect of RE-1 on UCN nonspecific adhesion. **a** is the fluorescence microscope picture, showing the field of view of the whole hole of the 96-well plate with A magnification of 100. **b** is the quantitative statistical results of Figure **a**.

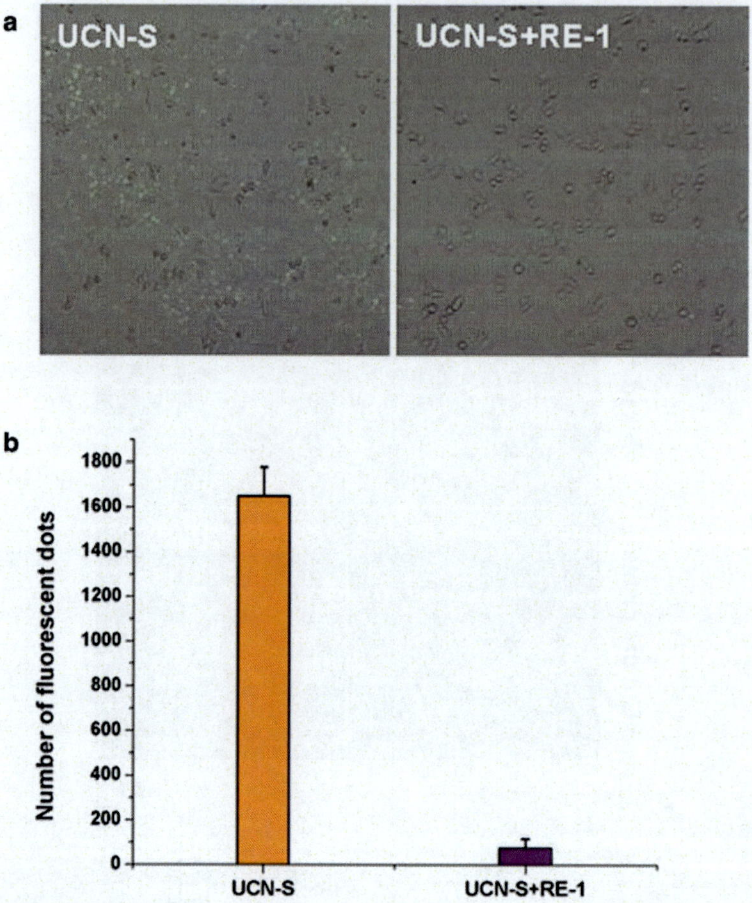

Fig. 4.5 Identification of the effect of RE-1 on UCN-S nonspecific adhesion on the surface of living cells. **a** is the fluorescence microscope picture, and **b** is the quantitative statistical result of Figure **a**

the change of deposition capacity of rare earth up-conversion luminescent nanomaterials coated with RE-1 short peptide (method is detailed in Sect. 4.4.5). In order to obtain more reliable data, we introduced PEG molecules widely used to improve the non-specific adhesion and dispersion of inorganic nanomaterials as the control. Through analyzing the detection results of time dynamics (as shown in Fig. 4.6a), we found that RE-1 short peptide could effectively reduce the settling rate of UCN and make UCN have good suspension capacity. High concentration PEG only showed some improvement in UCN suspension.

Force, but still much less effective than the RE-1 short peptide; RE-1 analogues (control peptides) under the same conditions did not. At the same time, we also found that for UCN-S nanomaterials with small particles, only a small amount of

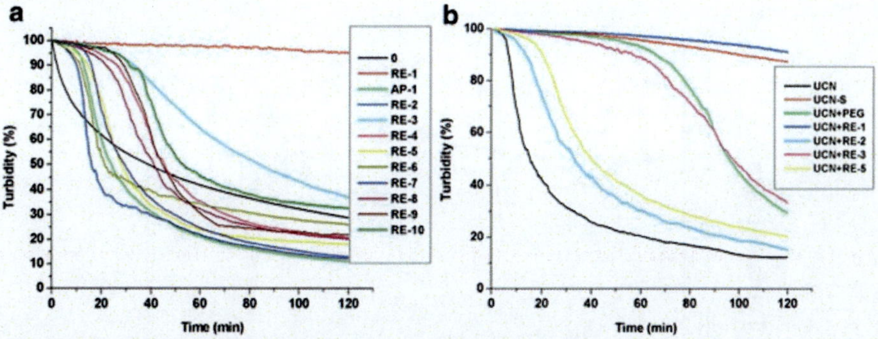

Fig. 4.6 Time dynamics curve. **a** is UCN, and **b** is UCP

sedimentation capacity was shown. The same method also detected UCP nanomaterials with large particles and obtained the same experimental results as UCN: RE-1 short peptide can effectively reduce the settling rate of UCP and improve the suspension capacity of UCP. The RE-1 analogue and control peptide AP-1 did not have this effect (as shown in Fig. 4.6b). In order to exclude that the above experimental results were caused by PEG not adsorbed on the surface of UCN, we further identified the interaction between UCN and PEG to ensure the accuracy of the experimental results. We first used ^{1}H NMR to detect the non-specific adsorption PEG on the surface of UCN. PEG (10 mg/mL) was incubated with UCN (100 μg/mL) according to the binding method, and PEG (unbound) in the product was washed off with NMR buffer, and the NMR one-dimensional hydrogen spectrum was scanned with NMR spectrometer at 298 K. As shown in Fig. 4.7, the binding product of UCN and PEG has a strong PEG characteristic peak at around 3.5 ppm, indicating that PEG molecules are indeed adsorbed on the surface of UCN. Then FTIR infrared spectroscopy was used to further verify this conclusion. As shown in Fig. 4.8, the binding products of UCN and PEG showed all the characteristic absorption peaks

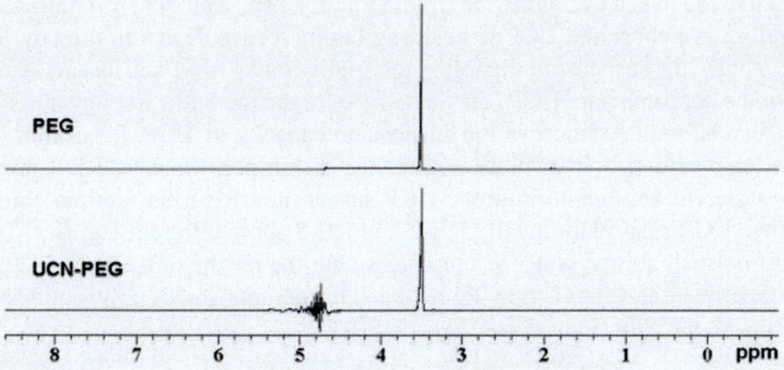

Fig. 4.7 Verification of PEG adsorbed on UCN surface by ^{1}H NMR

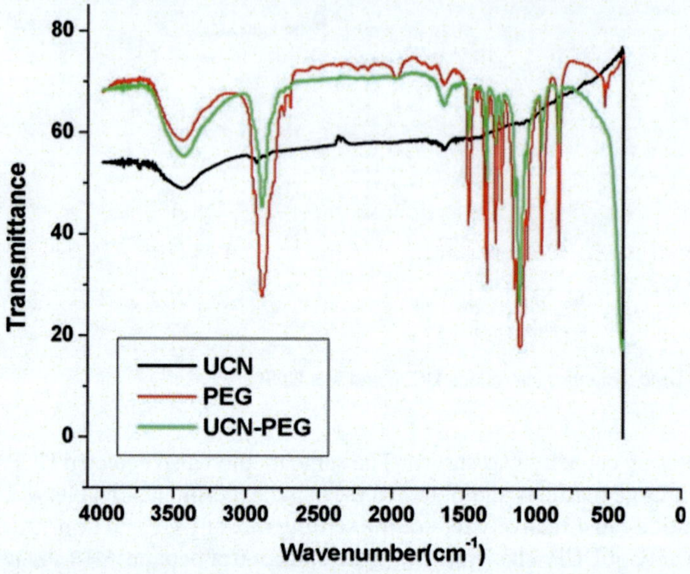

Fig. 4.8 FTIR verifies the PEG adsorbed on UCN surface

of PEG, indicating that PEG was bound to the surface of UCN. H NMR and FTIR infrared spectra showed that PEG was indeed adsorbed to the surface of UCN.

4.5.4 RE-1 Enhances the Suspension Capability of Rare Earth Up-Conversion Luminescent Nanomaterials

Due to the unique up-conversion luminescence characteristics of rare earth up-conversion luminescence nanomaterials, the green fluorescence of the materials can be seen to the naked eye under the irradiation of a hand-held 980 mm infrared laser. Therefore, we chose the UCP with strong luminescence design to directly observe and compare the suspension capability test of the coated RE-1 and its analogues with that of the uncoated pure UCP (see Sect. 4.4.6 for the method). As shown in Fig. 4.9, RE-1 can effectively improve the suspension capacity of UCP. Even after 24 h at room temperature, it still shows strong suspension performance. RE-1 analogues show different abilities to improve UCP suspension, RE-3 has certain abilities to improve UCP suspension, while RE-2 and RE-5 show different abilities. There is almost no such ability, which is consistent with the results of the peptide competition experiment in Chap. 2 (see Table 1) and the comparison of polypeptide binding ability with UCN in Chap. 3 (see Sect. 3.5.9).

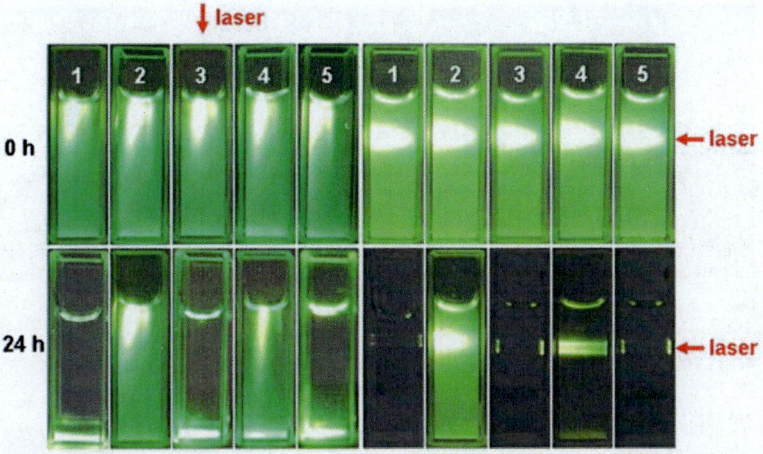

Fig. 4.9 Up-conversion fluorescence comparison of suspension capacity before and after 24 h. 1 is UCP, 2 is UCP + RE-1, 3 is UCP + RE-2, 4 is UCP + RE-3, and 5 is UCP + RE-5. The arrow points in the direction of 980 mm laser irradiation

4.5.5 RE-1 Reduces the Diffusion Capacity of Rare Earth Up-Conversion Luminescent Nanomaterials

In order to further identify the effect of RE-1 on the diffusion capacity of rare earth up-conversion luminescent nanomaterials. We modified the device in article, such as Eun Chul Cho [1], and designed a device that can effectively detect the diffusion capacity of the up-conversion luminescent nanomaterials of rare earth (see Fig. 4.2). The device can detect the interaction between cells and nanomaterials when cells are facing down and cells are facing up, respectively. When the cell is facing up, the nanomaterials in the solution need to reach the cell surface through sedimentation and diffusion. When the cell is facing down, the deposition ability of the nanomaterials does not work, but only reaches the cell surface through diffusion. In this way, the sedimentation and diffusion capacity of nanomaterials can be assessed by fluorescence microscopy with a 980 mm red external laser to detect the amount of nanomaterials interacting with cells. The result, as expected, was that for uncoated rare earth up-conversion luminescent nanomaterials, the amount of nanomaterials that nonspecifically interact with the cell surface when the cell is facing up is much higher than the amount of nanomaterials that interact with the cell surface when the cell is facing down. This is due to the synergistic effect of the sedimentation and diffusion capacities of nanomaterials (Figs. 4.10 and 4.11).For the package of the RE-1 on rare earth conversion luminescent nanomaterials, whether cells up or down, RE-1 were greatly reduced on rare earth conversion luminescent nanomaterials interact with cells of nonspecific, RE-1 that not only can reduce the sedimentation rate of rare earth on conversion luminescent nanomaterials, but also can reduce the diffusion ability, greatly enhance the stability of nanometer materials in solution.

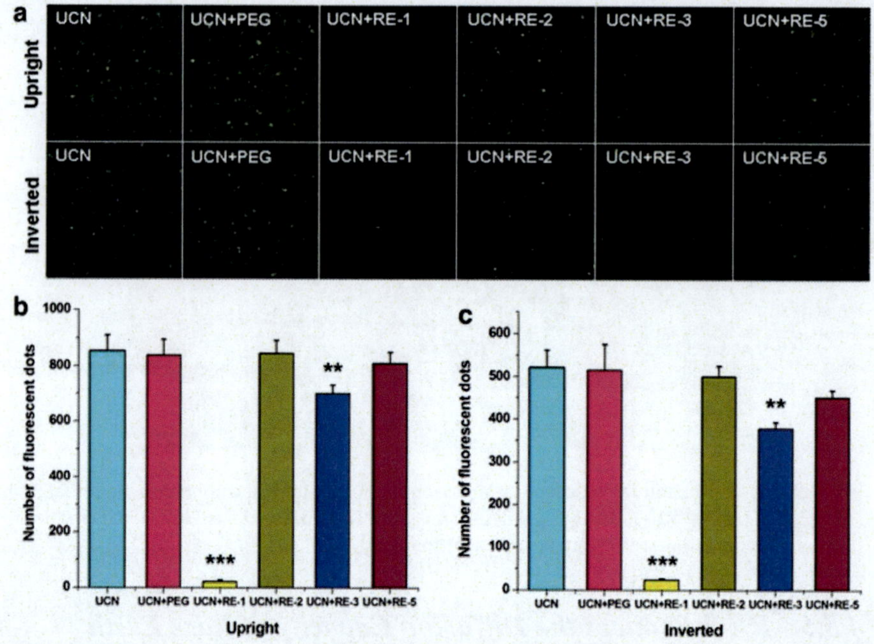

Fig. 4.10 Fluorescence microscopy to identify the diffusion capacity of UCN with cells facing up and down. **a** is the fluorescence microscope image of UCN that directly observed green fluorescence, **b** is the quantitative statistical diagram of UCN green fluorescence points when the cell was facing up, and **c** is the quantitative statistical diagram of UCN green fluorescence points when the cell was facing down. The magnification is 100

We also examined the effects of RE-1 analogues and PEG on the diffusion capacity of rare earth up-conversion luminescent nanomaterials. Consistent with previous experimental results, RE-3 showed some activity in reducing the diffusion capacity of rare earth up-conversion luminescent nanomaterials, while RE-2 and RE-5 had almost no such ability. Although the time moves at the above Sect. 4.5.3. In the mechanical experiment, PEG improved the suspension ability of UCN, but in this experiment, PEG hardly affected the interaction between rare earth up-conversion luminescent nanomaterials and cells, whether in the cell up-conversion group or the cell down-conversion group (as shown in Figs. 4.10 and 4.11).

4.5.6 RE-1 Can also Reduce the Non-Specific Interaction Between UCN and Cover Glass

Using device 4.2, we also examined the effect of RE-1 on non-specific interactions between UCN and cover glass without cultured cells. We also tested the amount of UCN adhered to the front and back sides of the cover glass (one side downward and

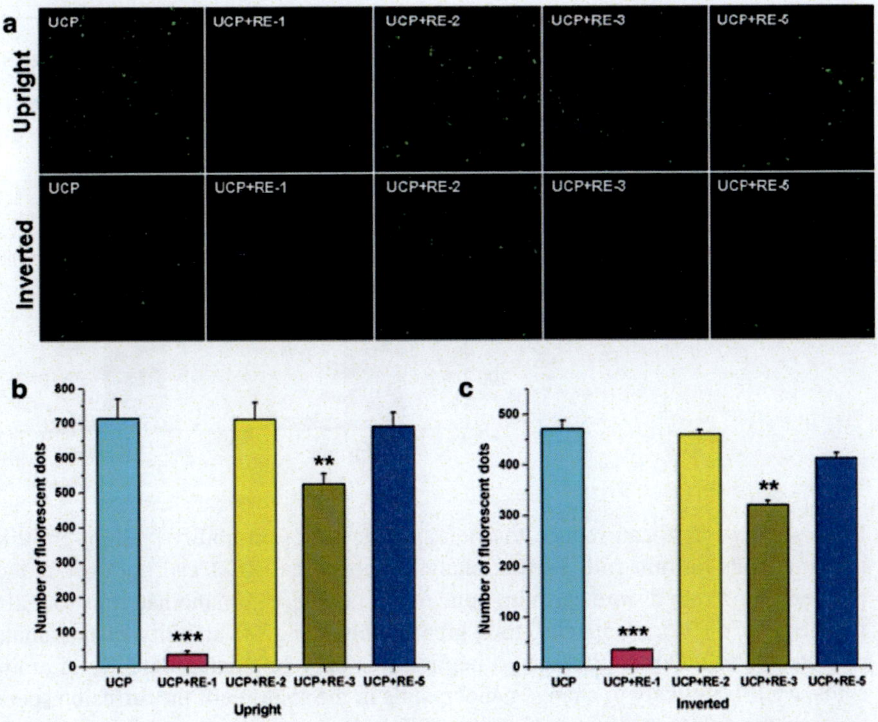

Fig. 4.11 Fluorescence microscopy to identify the diffusion capacity of UCP with cells facing up and down. **a** is the fluorescence microscope image of UCP that directly observed green fluorescence, **b** is the quantitative statistical diagram of UCP green fluorescence points when the cell faces up, and **c** is the quantitative statistical diagram of UCP green fluorescence points when the cell faces down. The magnification is 100

one side upward), and the results were consistent with the previous experiment. RE-1 could also reduce the non-specific interaction between UCN and the cover glass. For uncoated pure UCN, the adhesive amount on the upper surface of cover glass is much higher than that on the lower surface. In the UCN sample group coated with re-1, the UCN on both sides of the cover glass was greatly reduced (as shown in Fig. 4.12), further proving that RE-1 short peptide can reduce the non-specific adhesion of UCN to the surface of biological and non-biological media, enhance the SNR of UCN, and promote the application of UCN.

4.6 Summary

In this chapter, it is proved that the optical properties of rare earth up-conversion luminescent nanomaterials will not be affected by the inclusion of RE-1 short peptide, which is a prerequisite for the application. At the same time, we also found that

Fig. 4.12 RE-1 reduces the adhesion of UCN to cover glass slides

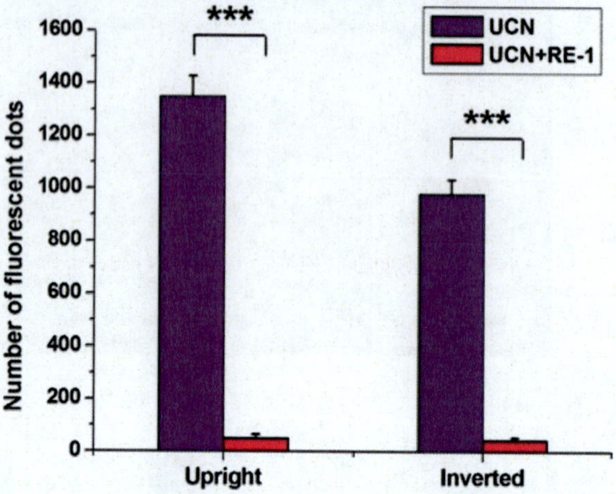

RE-1 short peptide can reduce the non-specific adhesion ability of light-emitting nanomaterials on thin soil. Further analysis shows that RE-1 can not only reduce the deposition rate of rare earth up-conversion luminescent nanomaterials, but also reduce their diffusion capacity, thus greatly enhancing the stability of nanomaterials in solution. Due to RE-1 short peptide was only 11 small molecules of amino acids, itself is unlikely to change nanocrystals in the solution of the diffusion speed, through the analysis, we conclude that RE-1 short peptide wrapped in rare earth nanomaterials of nanocrystalline surface formation of the peptide coating has two significant effect: first, stop the nanocrystalline precipitates, improved suspension. This effect may be achieved by reducing the interaction between materials. Second, reduce the nonspecific interaction between nanocrystals and cells. The first effect prevents the nanomaterials from reaching the surface of the cellular, non-biological medium, while the second effect drops. Once the nanomaterials have reached the cell surface through diffusion and sedimentation, they interact nonspecifically with cells or abiotic media. In living cells, RE-1 peptide may also reduce endocytosis.

Has reported material specific binding peptides most don't have to improve the performance of the nano materials in water suspension ability, this article found that the rare earth nanomaterials is of great application value and specific binding peptides, don't to do not need to enter the cells play a role of nano material and devices, we can block or artificial control, application in order to achieve the purpose.

Reference

1. Cho, E. C., Zhang, Q., & Xia, Y. N. (2011). The effect of sedimentation and diffusion on cellular uptake of gold nanoparticles. *Nature Nanotechnol, 6*(6), 385–391.

Chapter 5
RE-1 Effectively Shields the Cell Self-Effect of Rare Earth Nanomaterials, Reduces Their Toxicity and Improves Their Biological Safety

5.1 Introduction

With the continuous development of the application of nanomaterials in the field of biomedicine, the study of the biological effects of nanomaterials has also become a hot spot in this field. As an important cellular biological effect of nanomaterials, cell autophagy has been paid more and more attention worldwide to study its process and effect. Studies have shown that many inorganic and organic nanomaterials can trigger autophagy in a variety of cells, so increasing the level of autophagy may be a universal response of cells to nanomaterials. However, numerous studies have shown that autophagy is a "double-edged sword". Autophagy triggered by nanomaterials mostly promotes cell death, which further deepens people's concerns about the safety of nanomaterials. In the development and application of nanomaterials, the phenomenon of autophagy caused by nanomaterials cannot be ignored as a safety issue. In this chapter, the ability of rare earth nanomaterials to induce autophagy in vivo and in vitro was evaluated, and the effects of RE-1 short peptide encapsulation on the autophagy induction and cytotoxicity of rare earth nanomaterials in vivo and in vitro were investigated.

5.2 Experimental Materials

5.2.1 Reagents

LC3 antibody (NB100-2220, Novus, USA)
GFP antibody (sc-9996, Santa Cruz Biotechnology, USA)
P62 antibody (BML-PW9860-0100, BioMol, USA)
GAPDH antibody (ab8245, Abcam, USA)

Y. Zhang, *Tuning Autophagy-Inducing Activity and Toxicity for Lanthanide Nanocrystals*, Springer Theses, https://doi.org/10.1007/978-981-16-8166-0_5

Rabbit anti-mouse red fluorescent secondary antibody (sc362279, Santa Cruz Biotechnology, USA)
Transfection reagent Lipofectamine 2000 (11,668,019, Invitrogen, USA)
Bafilomycin A1 (023–11,641, Wako, Japan)
3-MA (3-methyl adenine) (M9281, Sigma, USA)
G418 (V7983, Promega, USA)
Bibenzimidazole dye (Hoechst 33,342, HO) (B2261, Sigma, USA)
ECL Development Kit (EZ-ECL, Biological Industries, Israel)
Protein marker (94KD), cell lysate and powerful tissue lysate were all purchased from China Beyotime Biotechnology Research Institute.
Medium powder, serum and trypsin used for cell culture were purchased from Invitrogen Company.

5.2.2 Cell Line

HeLa cells: human cervical cancer cell line (purchased from Shanghai Institute of Biochemistry, Chinese Academy of Sciences)

GFP-LC3/HeLa cells: cell lines stably transfected with eGFP-LC3 plasmid, in which eGFP-LC3 plasmid was donated by Professor Noboru Mizushima, Tokyo Metropolitan Institute of Clinical Medicine, Japan.

5.2.3 Experimental Animal

Balb/c mice (purchased from Shanghai Slax Animal Center, Shanghai, China) were raised in SPF grade sterile animal rooms.

5.2.4 Experimental Device and Consumables

Tissue grinder (Tenbroeck, WHEATON, USA)
Nitrocellulose membrane (Millipore, USA)
Slide (Beijing Zhongshan, China)
Cell culture dishes and cell culture plates related to cell experiments were purchased from Corning Company, USA.

5.3 Instruments and Equipment

Fluorescence microscopy (Olympus IX71, Olympus, Japan)
Vortex mixer (MS1 Minishaker, IKA, China)
Microcentrifuge (Centrifuge 5415D, Eppendorf, Germany)
Cell incubator (Scientific 3110, Thermo, USA)
Electrophoresis apparatus (Tanon EPS300, TANON, China)
Transmission electron microscopy (JEOL-1230, JEOL, Japan)

5.4 Experimental Methods

5.4.1 Cell Culture Method

All cells were cultured in a 37 C cell incubator containing 5% CO_2 in a modified Eagle's medium (DMEM medium) supplemented with 10% fetal bovine serum, 100 U/ml penicillin and 100 μg/ml streptomycin.

5.4.2 Establishment of GFP-LC3/HeLa Cell Line Stably Expressing GFP-LC3

HeLa cells in good condition and few passages were cultured in six-well plates, and the fusion degree of cells reached about 90% after 24 h. Add 1 μg eGFP-LC3 plasmid into 50 μL serum-free antibiotic-free double-free medium and mix evenly. At the same time, dilute 6 μL Lipofectamine 2000 with 50 μL serum-free antibiotic-free double-free medium, and mix evenly. After 5 min, the above-mentioned plasmids and transfection reagents were mixed together, and were added into the cells in six-well plates after 30 min at room temperature. The volume of medium in the wells was supplemented to 1 mL with double-free medium. After 6 h, fresh normal DMEM medium was added and the cells were cultured in a 37 C cell incubator. After 24 h, the cells were digested with trypsin and screened in DMEM medium containing 0.6 mg/ml G418, which was replaced every three days (DMEM containing 0.6 mg/ml G418). During the screening process, cells not transfected with G418-resistant plasmids would slowly apoptosis under the action of antibiotics, while there would be more and more GFP-LC3 fluorescent cells containing the G418-resistant plasmids. Ten days later, monoclonal selection was performed under fluorescence microscopy. Monoclonal cells with green burning light were selected and digested by trypsin and cultured in 0.3 mg/ml G418 screening solution alone. The cells screened by this method have the ability to stably express GFP-LC3, named as GFP-LC3/HeLa cells.

5.4.3 Experimental Methods of Inducing Cell Autophagy

GFP-LC3/HeLa cells were seeded in 96-well cell culture plates with a density of about 1~ 2 × 10^4 cells per well (if it was a 24-well cell culture plate, the density of GFP-LC3/HeLa cells was about 5 × 10^4/well) and cultured overnight. The next day, samples can be added when the cell fusion degree reaches 90%. Before the addition of samples, the cells were first replaced with fresh DMEM medium, and then the samples to be tested (depending on the experimental requirements) were added into the cells. At the same time, blank control (adding equal volume of PBS) and positive control (the concentration of autophagy inducer trehalose was 100 mM) were set up. The cells were treated in 37 C cell incubator for 24 h and then observed under fluorescence microscopy.

5.4.4 Statistical Method of GFP-LC3 Punctate Aggregation Positive Cells

The autophagic GFP-LC3/HeLa cells were observed with a burst light microscope, and the green fluorescent GFP-LC3/HeLa cells could be observed under the microscope by choosing the 488 nm blue laser with the fluorescent microscope. There will be a large number of green fluorescent aggregation points in the autophagy cells, when the number of green fluorescent aggregation points is greater than 5, the cells are positive cells. The visual field was randomly selected for photographing, and the proportion of positive cells with more than 5 GFP-LC3 punctate aggregates per 500 cells in total cells was calculated.

5.4.5 Experimental Method of Self-Accompanying Marker Staining

According to method 5.4.3, GFP-LC3/HeLa cells were treated with UCN at a final concentration of 1 mg/ml for 24 h, and stained with 10 μM dansylpentadiamine (MDC), 20 nM mitochondrial red fluorescent probe (MT), 75 μM acid lysosomal red fluorescent probe (LT) for 15 min, respectively. Then the cells were washed twice with PBS and the cell culture plate was placed directly under the fluorescence microscope for observation. Under the same field of view, green fluorescence of GFP-LC3 can be observed under blue laser excitation, and red fluorescence of MT or LT can be observed under green laser excitation. Then the two pictures are superimposed separately by image processing software.

5.4.6 Methods of Inducing Autophagy in Liver Tissue

Samples were prepared by 3.4.4 peptide binding method: uncapsulated UCN, UCN bound to 10 mg PEG, UCN encapsulated RE-1, uncapsulated UCN, UCP encapsulated RE-1 and UCP encapsulated AP-1.Balb/c mice were injected with the above prepared samples via tail vein at a dose of 15 mg/kg. A blank control mouse was set up and only injected with the same volume of saline. After 24 h, the liver tissues of each mouse were taken out and divided into two small and one large triplicates, two of which were used for tissue sample preparation and paraffin section preparation for transmission electron microscopy observation, and the other one for Western Blot detection.

5.4.7 Western Blot Detection Method

5.4.7.1 Processing Methods for Cell Samples

According to method 5.4.3, GFP-LC3/HeLa cells were seeded in 24-well cell culture plates and treated with the samples to be tested for 24 h, then the cells were collected by trypsin digestion. Add 40 μL of cell lysate containing protease inhibitor and 10 μL of 5 × SDS electrophoresis sample loading buffer, mix evenly, boil the sample in boiling water for 10 min, then put it into the microcentrifuge at the highest speed for 2 min to centrifuge the sample volatilized by boiling on the tube wall, and gently flick the bottom of the centrifuge tube to mix it.

5.4.7.2 BCA Protein Quantification

According to the number of samples, add 50 volumes of BCA reagent A to 1 volume of BCA reagent B (50:1) to prepare appropriate amount of BCA working solution, and fully mix it for reserve (BCA working solution is stable within 24 h at room temperature). Completely soluble protein standard BSA (concentration is 5 mg/ml), take 10 μL of water to dilute to 10 μL, so that the final concentration is 0.5 mg/ml (in what solution the protein sample is in, what solution the standard should be diluted with); at the same time, dilute the sample to be measured by three gradients of 10^{-1}, 10^{-2}, 10^{-3} for future testing. Take a clean 96-well plate, add 0, 1, 2, 4, 8, 12, 16, 20 μL of the diluted standard to the standard hole of the 96-well plate, add water to make up to 20 μL. 10 μL of sample to be tested diluted into gradient (each gradient is made into a compound hole) is added to the sample hole of 96-well plate, and water is added to make up to 20 μL. 200 μL of BCA working solution was added into each well and placed at 37 C for 30 min. The 96-well plate was removed and the A562 was determined by ELISA. Finally, a standard curve was made according to the standard and the protein concentration of the sample to be tested was calculated.

5.4.7.3 Methods for Processing Liver Tissue Samples

According to 5.4.6, the liver tissues taken from mice for Western Blot detection were cut with clean scissors, 1 ~ 2 mL of tissue lysate was added, and the homogenate was fully ground on ice with a tissue homogenizer until it turned transparent. The dissolved homogenate of liver tissue was centrifuged at 12,000 rpm for 5 min at 4 C, and the supernatant was transferred to a new centrifuge tube. 100 μL of the samples were taken out for testing, and the remaining samples were stored in a −20 C refrigerator for standby. In order to facilitate the detection of Western Blot, before detection, we used the above protein quantification method (see method 5.4.7.2) to adjust the protein concentration in each sample, took out 40 μL and added 10 μL of 5 × SDS electrophoresis sample-loading buffer, mixed them, boiled the sample in boiling water for 10 min, and then put it into the microcentrifuge at the highest speed for 2 min. Centrifuge the sample volatilized by boiling on the tube wall, and gently flick the bottom of the centrifuge tube to make it uniform.

5.4.7.4 Protein Electrophoresis

15% SDS-PAGE gel run electrophoresis of 15 holes was prepared by the following methods:

5 × electrophoresis buffer: 15.1 g Tris and 94 g Glycine were weighed, dissolved in 900 mL Milli-Q water, 50 ml of 10% (W/V) SDS was added, and Milli-Q water was added to 1 L at constant volume, ready for use.

5 × SDS-PAGE loading buffer: 1 M Tris–HCl (pH 8.8) 1.25 ml, SDS 0.5 g, BPB (bromate blue) 25 mg, glycerol 2.5 ml plus Milli-Q water to 5 mL for standby, when in use, add 25 μL beta mercaptoethanol to 500 μL, and then mix with vortex oscillation.

30% acrylamide storage solution: weigh 29.2 g acrylamide, 0.8 g methylene bisacrylamide and add 100 mL Milli-Q water to dissolve, then test its pH value, which should not be more than 7.0, and store in dark after filtration.

Prepare 5% concentrated gel and 15% separating gel for reserve.

	5% Concentrated gel	15% Separating gel
H_2O (mL)	3.4	1.7
30% acrylamide	0.83	4
1.5 M Tris–HCl pH6.8 (mL)	0.63	–
1.5 M Tris–HCl pH8.8 (mL)	–	2
10% SDS (μL)	50	80
10% APS (μL)	50	80
TEMED (μL)	5	3.2

Add 1 × electrophoresis buffer to the electrophoresis tank for sample addition. Before sampling, the above cell samples (or liver tissue samples) to be tested were briefly centrifuged at high speed, mixed and then sampled. At the same time, a pre-stained protein Marker was used to determine the electrophoresis time and analyze the electrophoresis results. The electrophoretic separation was performed under isovoltage conditions (80 V for the concentrated gel voltage and 120 V for the separated gel voltage).At the same time, observe the position of the 12 KB pre-stained Marker and the blue SDS loading buffer. When the blue loading buffer completely runs out of the gel (at this time, the 12 KB pre-stained Marker is at the bottom of the gel), stop electrophoresis.

5.4.7.5 Transmembrane

Preparation of 1 × electroporation buffer: weigh 3.02 g Tris (25 mM), 14.4 g Glycine (192 mM), add Milli-Q water 500 mL to dissolve, add 200 mL methanol, and finally set the volume to 1 L, ready to use. Cut filter paper and nitrocellulose membrane were immersed in a 20% ethanol-containing rotary buffer. The gelatin that has run the electrophoresis is cut off the upper gel of the comb and soaked in the electroporation buffer at the same time. In the rotating buffer, from cathode to anode, the splint box is clamped in the splint according to the order of "sponge—three-layer filter paper—protein electrophoresis gel—NC membrane—three-layer filter paper—sponge". The splint box is placed in the transfer membrane groove, and then the transfer membrane groove and the cooling box are placed in the transfer cell. Finally, the cell was filled with electrolyte and the membrane was transferred with 350 mA current for 90 min. Note: all operations are performed in an ultrasound-assisted electroporation solution to avoid the formation of bubbles affecting protein transfer.

5.4.7.6 Blocking

Prepare 10 × Ponceau S dycing solution: weigh 5 g Ponceau S powder and dissolve in 25 ml glacial acetic acid, add Milli-Q water to 50 mL for standby. 1 × Ponceau S: dilute 10 × Ponceau S with water 10 times.

NC membranes electrotransferred to proteins were stained with Ponceau red dye to observe the approximate band pattern of proteins on the membranes and the location of the required proteins. According to the known molecular weight of the protein to be tested, the required proteins were cut off separately from the position of the control pre-stained protein Marker for incubation of different antibodies. TBST was used to gently wash off the staining of Ponceau S and the NC membranes were sealed with 5% skimmed milk powder for 2 h at room temperature.

5.4.7.7 Antibody Incubation

Prepare 1 × TBST washing solution: weigh Tris 2.4228 g, NaCl 8.875 g and Milli-Q water to 1 L, then add 1‰ Tween-20.

On the sealed NC film of the sample to be tested, the primary antibody was dripped evenly and placed in a wet box at 4 C overnight. The next day, the NC membrane was removed and washed 5 times with 1 × TBST for 10 min each time. Then the secondary antibody labeled with horseradish peroxidase was incubated on a room temperature shaker with gentle shaking for 1 h, and then washed 5 times with 1 × TBST for 10 min each time.

5.4.7.8 ECL Development

In a darkroom, the above NC film was first dripped with horseradish peroxidase substrate (ECL) for color development, and at the same time, 1 × developing solution and fixative solution were respectively put into the plastic disk. The X-ray film was taken out under the red light, the X-ray film clip was opened, and the X-ray film was gently placed on the NC film. After 5 min, the exposed negative film was quickly immersed in the developing solution for development, and the fixative solution could be immersed after obvious strips appeared. Note: The exposure time can be adjusted according to the amount of protein and the titer of the antibody.

5.4.7.9 Western Blot Data Processing

In Western blot results, the size of protein signal band represents the molecular weight of protein, while the intensity of signal represents the expression level of protein. In order to better reflect the experimental results, we use the values obtained by comparing the experimental group with the internal participants to reflect the relative change trend of the experimental group. At the same time, we use Photoshop software to calculate the gray value of protein bands, and then we can count the change trend of protein expression level.

5.4.8 Transmission Electron Microscope (TEM) Biological Sample Observation Method

5.4.8.1 Cell Samples

GFP-LC3/HeLa cells were seeded in 24-well cell culture plates with a density of about 5×10^4 cells per well and cultured overnight for future use. Samples were prepared by 3.4.4 peptide binding method: uncapsulated UCN, UCN encapsulated RE-1, uncapsulated UCP, and UCP encapsulated RE-1. The blank control group was equal volume of PBS. According to the method 5.4.3, the cells were treated with the above samples (the final concentration of the materials were all 100 μg/ml) for 24 h. The cells were collected in a centrifuge tube of 1.5 ml, centrifuged at 5,000 rpm for

3 min to discard the culture medium, washed with PBS once, centrifuged to discard the supernatant, slowly added 1 ml of 2% glutaraldehyde dimethylarsine buffer (pH 7.4) along the wall of the centrifuge tube, and fixed in situ at 4 C for 6 h. Cell samples were then immobilized with 2% osmium tetroxide (OsO_4) at room temperature for 1 h, dehydrated with an ethanol gradient, and embedded in epoxy resin. The embedded cell samples were cut into ultrathin sections. After double staining with 3% uranium acetate-lead citrate, the sections were observed by transmission electron microscopy and photographed.

5.4.8.2 Liver Tissue Samples

The liver tissue prepared for TEM sample preparation in method 5.4.6 was quickly put into a centrifuge tube filled with 4 ml of 2% glutaraldehyde dimethylarsine buffer (pH 7.4) at 4 C for 12 h while the tissue (volume less than 1 mm^3) was removed. The method was the same as cell samples, fixed with 2% osmium tetroxide (OsO_4), then dehydrated with ethanol gradient and embedded in epoxy resin. The embedded tissue samples were cut into ultrathin sections. After double staining with 3% uranium acetate-lead citrate, the sections were observed by transmission electron microscopy and photographed.

5.4.9 MTT Colorimetric Assay for Cell Viability

Samples were prepared by 3.4.4 peptide binding method: uncapsulated UCN, UCN encapsulated with RE-1 and its analogues, UCN encapsulated with 10 mg PEG, uncapsulated UCP, UCP encapsulated with RE-1 and its analogues. According to method 5.4.3, after the cells were treated with the above samples for 24 h, 10 μL MTT solution (5 mg/ml) was added to each well, and after 4 h of further culture, the medium was carefully absorbed and 150 μL dimethyl sulfoxide (DMSO) was added to each well, which was placed on a shaker at room temperature and shaking for 10 min at low speed. After the crystals were fully dissolved, the absorbance values of each hole were measured at OD490 nm in an enzyme label instrument.

Note: The whole process of MTT experiment ensures aseptic operation, because bacteria can also cause the increase of MTT colorimetric OD value, which will affect the experimental results.

5.4.10 Cell Death Assay (PI/Hoechst Staining)

Samples were prepared by 3.4.4 peptide binding method: uncapsulated UCN, UCN encapsulated with RE-1 and its analogues, UCN encapsulated with 10 mg PEG, uncapsulated UCN, UCP encapsulated with RE-1 and its analogues, respectively.

According to method 5.4.3, cells at the above samples were stained with propidium iodide (PI) at a final concentration of 10 μg/ml and bisbenzimidazole dye (Hoechst 33,342) at the same final concentration of 10 μg/ml for 10 min after 24 h. Then it was observed under a light-changing microscope: PI could stain dead cells red under green light excitation; Hoechst could stain nuclei blue under ultraviolet light excitation to count the total number of cells. Finally, PI stained cells were divided by the total number of cells (counting more than 500 cells) and the proportional statistical analysis was performed to calculate the cell death rate.

5.4.11 Method of Making Paraffin Sections

Tissues prepared for paraffin sectioning in Method 5.4.6 were quickly removed and fixed in a centrifuge tube filled with 10 ml of 2% glutaraldehyde dimethylarsine fixative (pH 7.4) for 24 h at 4 C. Fixed tissue samples were repaired to a volume of $5 \times 5 \times 2\ mm^3$ and dehydrated with gradient ethanol: 70% ethanol was dehydrated three times, each time for 15 min; 85% ethanol was dehydrated for 15 min; 95% ethanol was dehydrated for 15 min; 100% ethanol was dehydrated three times, each time for 15 min. The sample was transparent twice with ethanol-xylene mixture (1:1 volume ratio) for 5 min each time, and then replaced with pure xylene transparent solution to continue transparent twice for 5 min each time. After the paraffin is heated and melted, it is put into the oven for 2–4 h, so that the temperature of paraffin is balanced at 52–58 C. Transparent tissue samples are put into paraffin for 3 times, 30 min each time, and finally embedded in paraffin to make embedded paraffin blocks. Paraffin sections (about 2 m) were cut with a microtome and attached to polylysine-treated slides, which were then dried in an oven at 37 C.

5.4.12 HE (Hematoxylin–eosin) Staining Method

The paraffin slices prepared in the above method 5.4.11 were put into the oven, dried at 60 C for 2 h, and dewaxed twice in dimethylbenzene heated to 40–60 C for 4 min each time. Then dehydrated with gradient ethanol, 100% ethanol for 1 min, 95% ethanol for 1 min, 75% ethanol for 1 min, soaked in Milli-Q water for 2 min, and dyed in hematoxylin dye for 5–10 min. Wash the excess dye off the slide with water and separate it with 0.5–1% hydrochloric alcohol for 10 s. Microscopic examination was controlled until the nucleus and chromatin in the nucleus were clear. Rinse with water for 15 min. Continue to immerse in 0.5% eosin staining solution for 5 min. The stained sections were then dehydrated by gradient ethanol, followed by 70, 85, 95 and 100% dehydration for 5 min at all levels to remove water from the tissues. It was then immersed in xylene transparently twice for 10 min each time. Finally, the excess dimethylbenzene around the slices was wiped off, and the slices were immediately sealed and put into a 37 C supply box for dry reserve.

5.4.13 Immunofluorescence Detection

The paraffin slices prepared in the above method 5.4.11 were put into the oven, dried at 60 C for 2 h, and dewaxed twice in dimethylbenzene heated to 40–60 C for 4 min each time. Then they were dehydrated with gradient ethanol, 100% ethanol for 1 min, 95% ethanol for 1 min, 75% ethanol for 1 min, Milli-Q water for 2 min, and then dipped in PBS twice, each time for 2 min. The slices were placed in 0.1 mol/L citric acid solution (pH 6.0) and heated in a microwave oven for 6 min until slightly boiling to conduct the antigen repair. The slices were maintained at low and medium firepower for 10 min, and cooled naturally at room temperature after stopping heating. After cooling for 20–30 min, PBS was soaked for 2 min, and 0.3% TritonX-100 was added to break the membrane for 10 min. Wash the slices with PBS twice, 4 min each time, to wash off the membrane breaking fluid. Wipe the PBS outside the sample with filter paper, drop 10% FBS blocking serum in the wet box and seal at 37 C for 30 min. Wipe off the blocking solution with filter paper and directly drop 1:100 LC3-anti into a wet box and incubate overnight at 4 C. The next day, the PBS was soaked five times for 3 min each time, and the primary antibody was washed away, and the PBS outside the sample was wiped off with filter paper. Continue to drop Rhodamine Fluorescent secondary antibody and incubate in a humidified box at room temperature for 1–1.5 h in the dark. At this time, staining can be observed under the fluorescence microscope quickly to determine the termination time. The secondary antibody was washed 3 times with PBS, and the PBS outside the specimen was wiped off with filter paper. Hoechst 33,342 was added to the sample and incubated for 2 min in the dark, which could stain the nuclei of hepatocytes and show blue prominence under UV light excitation. Hoechst 33,342 was washed 3 times with PBS and the PBS outside the specimen was wiped off with filter paper. Finally, the film was sealed with a sealing liquid containing an anti-fluorescence quencher and immediately observed under a light-burst microscope.

5.4.14 Data Analysis

All experiments were repeated at least three times. All data were expressed as mean S.E.M., Student's two-tailed t-test. Data analysis was processed by Two-tailed student's test in Origin data analysis software. Statistically significant criteria were $^{*}p < 0.05$, $^{**}p < 0.01$, $^{***}p < 0.005$ as significant differences.

5.5 Experimental Results and Discussion

5.5.1 *Rare Earth Upconversion Luminescent Materials Can Induce Cell Autophagy*

It has been reported that many nanomaterials can trigger autophagy, which has become a special biological effect of nanomaterials on the nanoscale. To test this view, we tested the ability of three rare earth upconversion luminescent nanomaterials to induce cell autophagy by detecting the most direct marker of autophagy occurrence (GFP-LC3 punctate aggregation). Normal untreated GFP-LC3/HeLa cells were observed by fluorescence microscopy, and green fluorescent protein was uniformly and diffusely distributed in the cytoplasm. When cells were stimulated by treatment, the marker protein LC3 (GFP-LC3) protein of autophagic vacuoles would accumulate on the autophagic vesicle membrane along with the formation of autophagic vacuoles, forming bright green GFP-LC3 punctate aggregation. This detection method can most directly detect the occurrence of autophagy by fluorescence microscopy observation.

As shown in Fig. 5.1, GFP-LC3/HeLa cells treated with 100 μg/ml rare earth upconversion luminescent nanomaterials (UCN, UCP, and UCN-S) for 24 h showed different degrees of autophagic effect. Unlike Trehalose, a chemical inducer of autophagy, UCN and UCN nanoparticles are capable of vacuolating cells while initiating autophagy, which is another cytotoxicity distinct from the autophagic

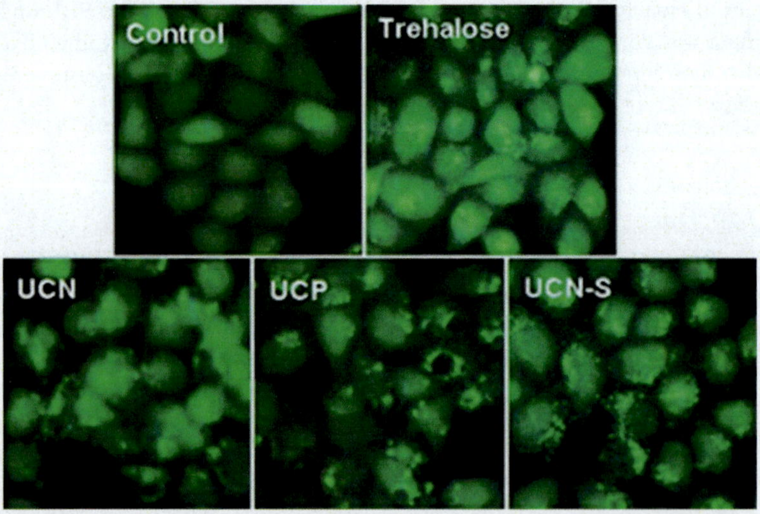

Fig. 5.1 Point aggregation of GFP-LC3 induced by rare earth upconversion luminescent nanomaterials. Control: untreated cells; Trehalose treatment concentration was 100 mM; UCN, UCP, UCN-S concentrations were 100 μg/ml

response of cells. In the figure, we observed that the cells treated with UCN and UCN group produced a large number of vacuoles, while UCN-S did not produce obvious vacuolization. This result is consistent with the conclusion that the rare earth nanomaterials induced cytoplasmic vacuolization has a size effect, and small particles are not easy to cause vacuolation as elucidated in the previous published articles of our group [1]. This result is also consistent with the recent report that gold nanoparticles can cause mitochondrial enlargement [2] and further cause cell cytoplasmic vacuolization.

The morphological detection method, which uses transmission electron microscopy to directly observe the microscopic structure of autophagosomes produced in cells, has been used as the gold standard to detect the occurrence of autophagy. Damaged organelles such as swelling and degeneration of mitochondria, vacuolar bilayer membrane-like structure appearing in cytoplasm, structure of bilayer membrane surrounding autophagosome and structure of autophagolysosome formed by fusion with lysosome can be observed under transmission electron microscopy. As shown in Fig. 5.2, in the untreated control group, the cells were in good condition, with normal cell structure and few vesicle structures in the cytoplasm, while after 24 h treatment with 100 μg/ml UCN, a large number of double-layer membrane-like autophagosome structures and autophagolysosome structures of monolayer membrane fused with autophagosome and lysosome can be observed under transmission electron microscopy, which further proves that UCN can induce the occurrence of cell autophagy.

Control was untreated cells; UCN was 100 μg/ml cells treated for 24 h. N represents the nucleus; the arrow points to an autophagosome with a double membrane structure inside the cell. Scale: 1 μm.

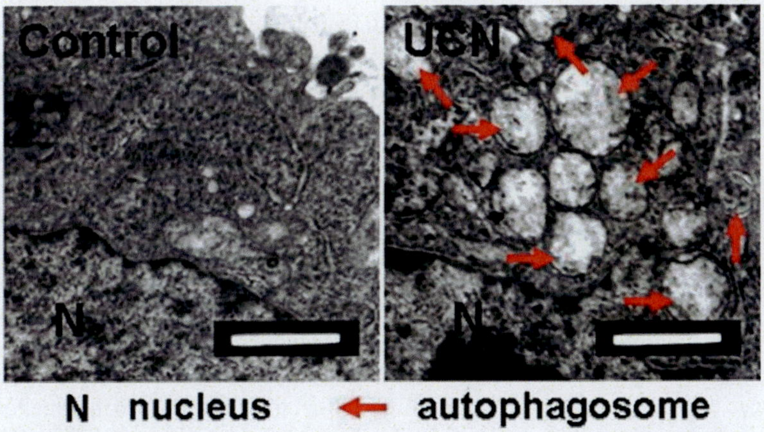

Fig. 5.2 Transmission electron microscopy of UCN-induced autophagy in HeLa cells

5.5.2 Cellular Autophagy Induced by Rare Earth Upconversion Luminescent Materials is a Complete Process

Cellular autophagy is a dynamic process, which includes not only the formation process of autophagosomes, the transport of autophagosome-encapsulated substrates to lysosomes for binding, but also the degradation and release of encapsulated substrates in lysosomes. Therefore, the detection of GFP-LC3 punctate aggregation only proves the formation of autophagosomes in the whole process of autophagy, which cannot be used as a sufficient basis for the occurrence of cell autophagy, and needs to be supplemented by other dynamic detection methods to fully prove it. To further determine the whole dynamic process of cell autophagy induced by rare earth upconversion luminescent materials, we next examined the process of encapsulating substrates and binding to lysosomes after autophagosome formation.

First, we examined the co-localization of GFP-LC3 protein with intracellular autophagy-related components. GFP-LC3/HeLa cells were treated with 100 μg/ml UCN for 24 h and stained with MDC, a stain for acidic vesicles, to observe its co-localization with GFP-LC3. DMC stain can stain the acidic components in the cells, but it is not acidic at the initial stage of autophagosome formation, while it will be acidic under the action of lysosomal acid phosphatase after the fusion of autophagosome and lysosome, so it can be stained by MDC stain. As shown in Fig. 5.3, the GFP-LC3 green fluorescent protein labeled autophagy co-localized with the red intracellular component after staining with acid dye MDC, which indicated that the fusion process between autophagy and lysosome occurred in GFP-LC3/HeLa cells after UCN treatment. Because when autophagy occurs, a large number of organelles such as mitochondria are wrapped in autophagosomes and autophagic lysosomes and degraded. Therefore, we stained mitochondria with the mitochondria-specific red fluorescent probe MT in living cells to observe its co-localization with GFP-LC3 green fluorescent protein, which labels autophagy. As shown in Fig. 5.3, there was obvious co-localization between GFP-LC3 green fluorescent protein and mitochondrial-specific red fluorescent MT, indicating that mitochondrial transport to autophagosomes or autophagolysosomes occurred in GFP-LC3/HeLa cells after UCN treatment. We further examined the fusion process between autophagosomes and lysosomes. LT is a lysosomal red fluorescent probe that can selectively reside in acidic lysosomes to achieve lysosome-specific fluorescent labeling in living cells. As shown in Fig. 5.3, GFP-LC3 green fluorescent protein labeled autophagosomes showed excellent co-localization with red lysosomes. This indicates that the formation of autophagosomes occurs in the treated GFP-LC3/HeLa cells, as well as the process of autophagosome encapsulating mitochondria and other organelles transporting to lysosomes and fusing with lysosomes, further demonstrating the ability of UCN to induce cell autophagy; at the same time, since autophagosomes have transported substrates to lysosomes and fused with lysosomes, and lysosomal acid phosphatase is active, thus it is also preliminarily demonstrated that the substrate degradation process of UCN-induced cell autophagy was complete.

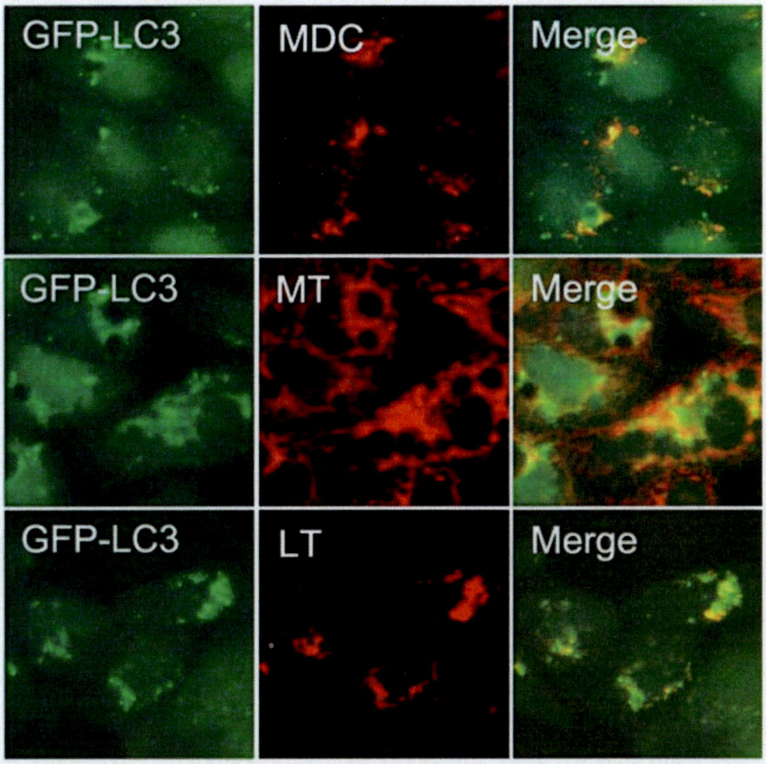

Fig. 5.3 Fluorescence staining to observe the co-localization of GFP-LC3 green fluorescent protein labeled autophagy and autophagy-related components

When autophagy occurs, the formation of autophagosomes gradually increases, and LC3-I proteins will be specifically cleaved into LC3-II and transferred to autophagosomes. LC3-II proteins are on the autophagosome membrane, and the autophagosome will eventually fuse with the lysosome and degrade within it. Therefore, the occurrence of autophagy can usually be determined by detecting the conversion of LC3-I to LC3-II and the level of LC3-II protein by Western Blot. However, the level of LC3-II protein can only reflect the number of autophagosomes, and the production of autophagosomes represents that autophagy is activated. However, the observation of more autophagosome aggregation does not necessarily mean that the induced autophagic activity is stronger. It may also be due to the accumulation of autophagosomes caused by the blockage of the downstream pathways of autophagy (such as the fusion of autophagosomes and lysosomes) after the formation of autophagosomes. Therefore, it is important to evaluate whether the autophagy induced by rare earth upconversion luminescent nanomaterials actually occurs in a complete autophagic flux.

To exclude that the increase of autophagosomes in UCN-treated cells may be the result of blockage of the downstream pathway of autophagy, we first employed the

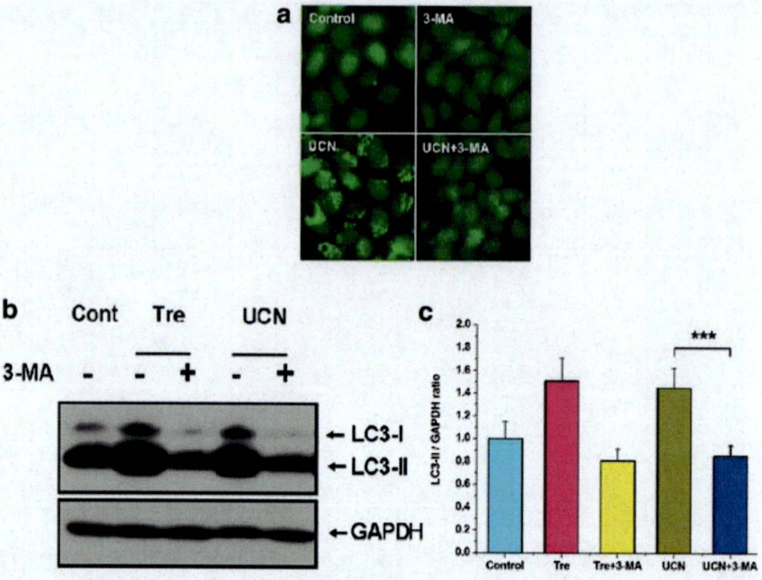

Fig. 5.4 Effect of 3-MA on induction of autophagy

autophagy-specific inhibitor 3-MA for treatment. GFP-LC3/HeLa cells were treated with 100 μg/ml UCN together with 5 mM 3-MA, and compared with 100 μg/ml UCN alone treatment group, it was found that 3-MA could inhibit the punctate aggregation of GFP-LC3 induced by UCN (Fig. 5.4a); Western Blot assay also showed that 3-MA could inhibit the conversion of LC3-I to LC3-II in UCN-treated cells, and also reduce the level of LC3-II protein in UCN-induced cells (as shown in Fig. 5.4b, c), which indicated that the increase of LC3-II protein level in UCN-induced cell autophagy was caused by the increase of autophagosomes (that is, the increase of LC3 protein synthesis). This result suggests that UCN-induced cell autophagy may be a complete autophagic flux.

Figure 5.4a shows the fluorescence microscopy images of GFP-LC3/HeLa cells treated with different treatments. Control is untreated cells, 3-MA is cells treated with 5 mM 3-MA alone, UCN is cells treated with 100 μg/ml UCN alone, UCN + 3-MA is cells treated with 100 μg/ml UCN and treated with 5 mM 3-MA together; Fig. 5.4b shows the changes of LC3-II protein levels of HeLa cells treated with different treatments, from left to right in the following order: untreated Control group, 100 mM trehalose alone treatment group, 100 mM trehalose and 5 mM 3-MA co-treatment group, 100 μg/ml UCN alone treatment group, 100 μg/ml UCN and 5 mM 3-MA co-treatment group; Fig. 5.4c is the quantitative statistical diagram of LC3-II/GAPDH of Fig. 5.4b, which further intuitively reflects the results of Fig. 5.4b and compares the LC3-II protein levels.

As mentioned above, the induction of autophagy is not only an increase in LC3-II or autophagosome formation, but more importantly, the detection of the dynamic process of the whole autophagic system, including lysosomes, as well as the integrity of the whole autophagic flow. Therefore, we further evaluated the fusion process between autophagy and lysosome in the downstream pathway of autophagy after activation of autophagy, i.e. detecting the level of LC3-II protein under the action of Bafilomycin A1, a lysosome fusion inhibitor. We added 400 nM Bafilomycin A1 to 100 μg/ml UCN-treated HeLa cells 4 h before sampling, and compared the LC3-II protein levels of UCN-treated HeLa cells without Bafilomycin A1. The results of Western Blot showed that the level of LC3-II protein in cells treated with UCN for 24 h was significantly higher than that in untreated control group; because Bafilomycin A1 could inhibit the fusion of autophagosome and lysosome, the degradation of LC3-II protein was blocked, resulting in the increase of LC3-II level, so the level of LC3-II protein in Bafilomycin A1 alone treatment group was also higher than that in the control group, while the level of LC3-II protein in UCN and Bafilomycin A1 co-treated cells was found to be higher than that in Bafilomycin A1 alone treatment group. This result showed that when the fusion of autophagosome and lysosome was blocked in the downstream pathway of autophagy, the level of LC3-II protein was still improved, indicating that UCN could induce more autophagy, that is, enhance the activity of cell-induced autophagy, further illustrating that UCN-induced cell autophagy is a complete process (Fig. 5.5).

The treatments in Fig. 5.5a from left to right were as follows: untreated control group, 400 nM Bafilomycin A1 alone treatment group, 100 mM trehalose alone treatment group, 100 mM trehalose and 400 nM Bafilomycin A1 co-treatment group, 100 μg/ml UCN alone treatment group, 100 μg/ml UCN and 400 nM Bafilomycin A1 co-treatment group (Bafilomycin A1 was added 4 h before sample collection); Fig. 5.5b is the quantitative statistical diagram of LC3-II/GAPDH in Fig. 5.5a. High exp. and Low exp. represent strong and weak exposures, respectively.

Next, we examined the degradation of the autophagic substrate P62. In order to better monitor the protein level of LC3-II and the degradation of the autophagic substrate P62 during the whole process, we chose to treat HeLa cells with 100 μg/ml UCN at different time points (0, 3, 6, 12, 18, 24 h), collected the cells, and then detected the conversion of LC3-I to LC3-II, the protein level of LC3-II and the

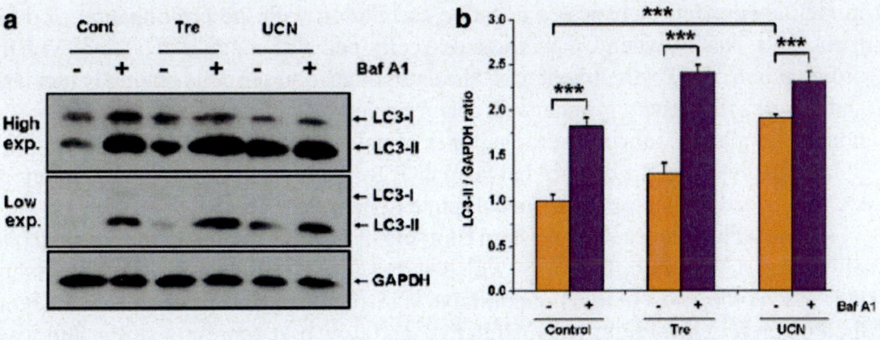

Fig. 5.5 Changes of LC3-II protein levels in UCN-treated cells by Bafilomycin A1

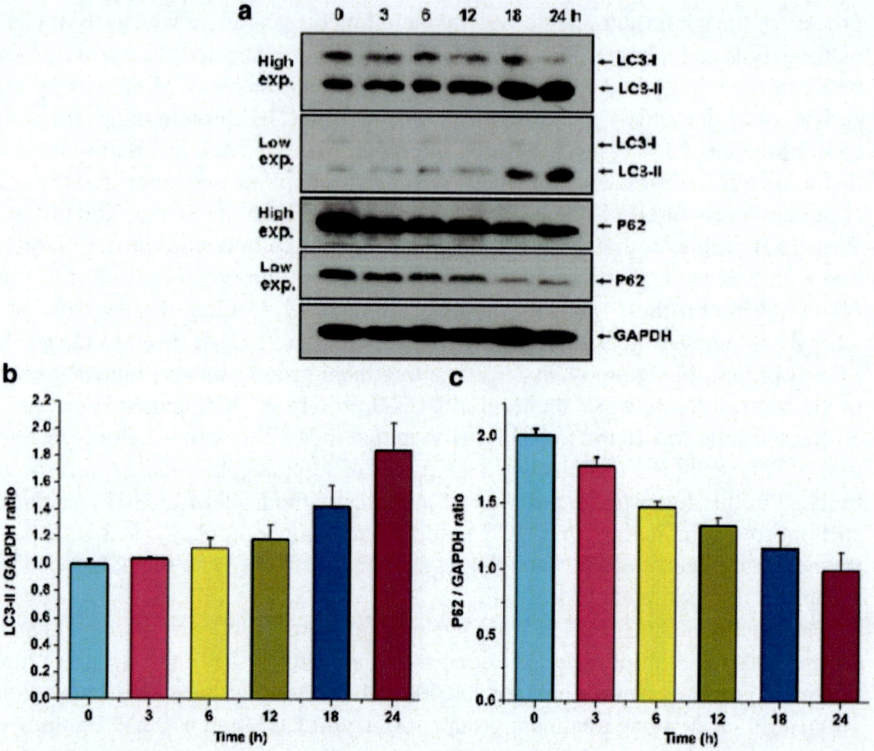

Fig. 5.6 Changes of LC3-II protein level and P62 protein level in HeLa cells after UCN treatment with time gradient

degradation of P62 protein by Western Blot with GAPDH as the internal reference. As shown in Fig. 5.6, UCN could activate the conversion of LC3-I to LC3-II in HeLa cells. The LC3-II protein level of HeLa cells treated with UCN increased significantly, and there was a positive correlation increase with the prolongation of UCN treatment time. In contrast, UCN treatment decreased the level of P62 protein in HeLa cells, and there was a negative correlation with the prolongation of treatment time. That is, when UCN is added to cells, cell autophagy is activated; with the prolongation of UCN treatment time, the autophagosomes in cells gradually increase, and autophagic activity is enhanced, accompanied by enhanced autophagic degradation of autophagic substrates; the number of autophagosomes observed at any time point is the result of the balance between their formation and degradation. Therefore, UCN-induced cell autophagy is a complete dynamic process.

Figure 5.6a is the result of Western Blot; Fig. 5.6b is the quantitative statistical map of LC3-ear/GAPDH in Fig. 5.6a, which shows the LC3-II protein level intuitively; Fig. 5.6c is the P62/GAPDH quantitative statistical map in Fig. 5.6a, which shows the P62 protein level intuitively. High exp. and Low exp. represent strong and weak exposures, respectively.

5.5.3 RE-1 Effectively Shields its Ability to Induce Cell Autophagy by Reducing the Interaction between Rare Earth Upconversion Luminescent Nanomaterials and Cells

Cellular autophagy is a "double-edged sword". Low-level autophagy is a necessary means to maintain the homeostasis of intracellular energy circulation and a self-protective mechanism of cells. When cells are stimulated by the external environment, cell autophagy will induce cell damage, leading to cell death, and most of the cell autophagy induced by nanomaterials is pro-cell death. Therefore, it is one of the key research contents in this paper to try to effectively control the autophagy induced by nanomaterials, especially rare earth upconversion luminescent nanomaterials widely used in biomedical field. So how can we effectively regulate the autophagic effect induced by rare earth upconversion luminescent nanomaterials? How do nanomaterials activate autophagic effects? To solve these problems, we tried to use transmission electron microscopy to visualize the UCN-treated cells and try to find the answer. We prepared samples of HeLa cells treated with 100 μg/ml UCN for TEM observation. The results showed that a large number of nanoparticle aggregates were found in the cells, and some nanoparticles were aggregated and encapsulated in the membrane structure (Fig. 5.7). This may be due to the fact that after the treatment of cells with UCN, they entered the cells by interacting with the cells and activated cell autophagy. So how do UCN nanoparticles enter cells? Except for small molecules that can enter the membrane through transmembrane channels, the exogenous substances of the cell almost exclusively enter the cell through endocytosis. To prove this point, we used the membrane-mediated endocytosis inhibitor genistein to inhibit the endocytosis of UCN. We co-treated HeLa cells with 100 μg/ml UCN and 50 μM genistein for 24 h and compared them with 100 μg/ml UCN alone treatment group. As shown in Fig. 5.8, in cells co-treated with UCN, genistein significantly inhibited the conversion of LC3-I to LC3-II caused by UCN and reduced the level of LC3-II protein.

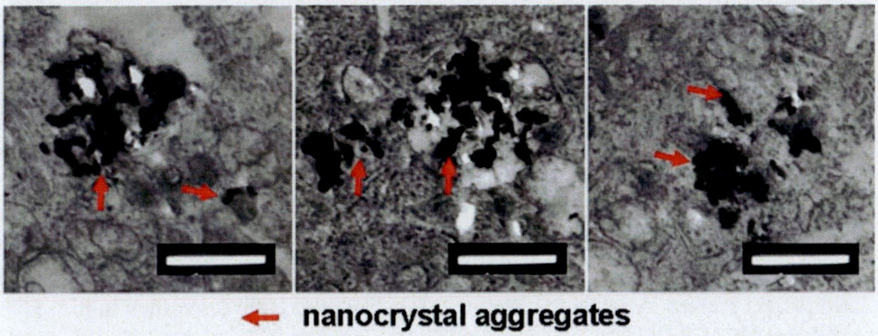

Fig. 5.7 Transmission electron microscopy of UCN-treated HeLa cells. Arrows point to large aggregates of nanoparticles within cells. Scale: 1 μm

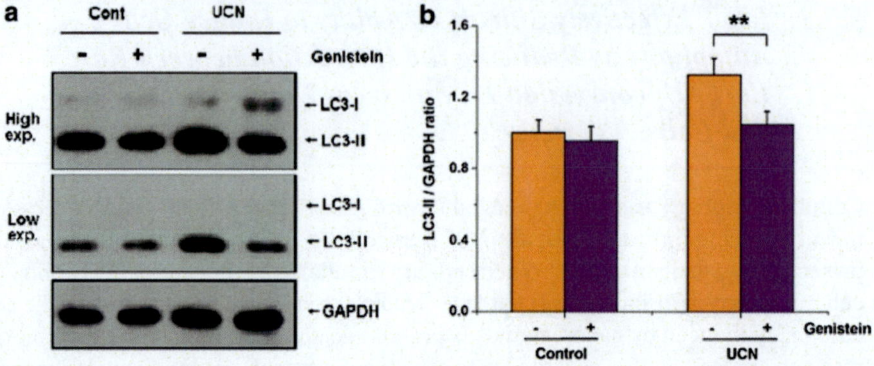

Fig. 5.8 Effect of genistein on LC3-II protein level of HeLa cells after UCN treatment

This result shows that genistein can inhibit the cell autophagy effect caused by UCN by inhibiting the endocytosis of UCN, further illustrating that UCN-induced cell autophagy can be achieved by reducing the non-specific interaction between UCN and cells.

Figure 5.8a is the result of Western Blot; Fig. 5.8b is the quantitative statistical graph of LC-3II/GAPDH in Fig. 5.8a, which visually shows the LC3-II protein levels. High exp. and Low exp. represent strong exposure and weak exposure, respectively.

In Chap. 4, it has been demonstrated that RE-1 short peptides can improve the suspension ability of rare earth nanomaterials in water and reduce the non-specific interaction between nanomaterials and cell and media surfaces. Therefore, using RE-1 to specifically bind short peptides to biologically modify rare earth upconversion luminescent nanomaterials may achieve effective regulation of the autophagic effect triggered by nanomaterials. Therefore, we compared the ability of RE-1-encapsulated and uncapsulated RE-1 upconversion luminescent nanomaterials to induce autophagy in GFP-LC3/HeLa cells by using fluorescence microscopy to observe the punctate aggregation of GFP-LC3. The punctate aggregation of GFP-LC3 in GFP-LC3/HeLa cells treated with different concentration gradients of 0, 10, 100, 1000 μg/ml UCN, UCP and UCN-S coated and uncoated by RE-1 peptides was observed. The results showed that the punctate aggregation ability of GFP-LC3 in cells induced by UCN, UCP and UCN-S had a dose effect, and the number and fluorescence intensity of punctate aggregation increased with the increase of material dosage, and showed a positive correlation trend; a large number of cytoplasmic vacuolation effects could be clearly seen in UCN and UCP treatment groups, and the intensity of vacuolization was also positively correlated with the dosage of added material; and RE-1 could significantly shield the punctate aggregation induced by the above three rare earth upconversion luminescent nanomaterials, and can effectively inhibit the cell vacuolation effect caused by UCN (as shown in Fig. 5.9) and UCP (as shown in Fig. 5.10). Although UCN-S did not cause cytoplasmic vacuolization, the punctate aggregation of GFP-LC3 of UCN-S was also significantly reduced after encapsulation of RE-1 (Fig. 5.11). This result preliminarily proves the previous hypothesis that RE-1

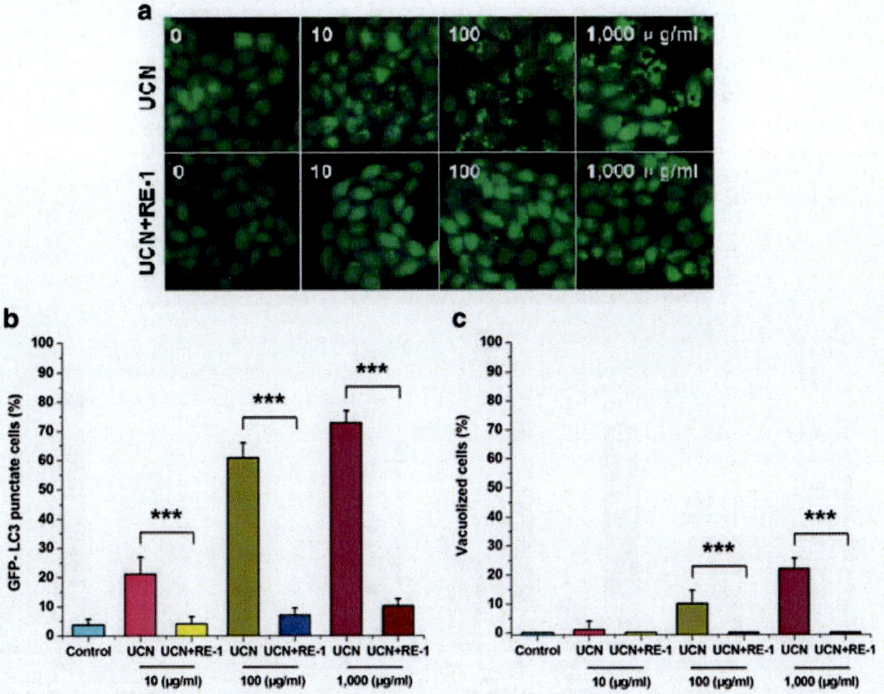

Fig. 5.9 RE-1 shields GFP-LC3/HeLa cells from punctate aggregation and cell vacuolization induced by UCN at different concentration gradients

short peptide can shield the autophagic effect induced by rare earth upconversion luminescent nanomaterials.

Figure 5.9a is a fluorescence microscope photograph, Fig. 5.9b is a quantitative statistical map of GFP-LC3 punctate aggregation in Fig. 5.9a, c is a quantitative statistical map of cell vacuolization in Fig. 5.9a.

Figure 5.10a is a fluorescence microscope photograph, Fig. 5.10b is a quantitative statistical map of GFP-LC3 punctate aggregation in Fig. 5.10a, c is a quantitative statistical map of cytoplasmic vacuolization in Fig. 5.10a.

Figure 5.11a is a fluorescence microscope photograph, and Fig. 5.11b is a quantitative statistical graph of GFP-LC3 punctate aggregation in Fig. 5.11a.

Next, we directly observed the changes of cell microstructure using transmission electron microscopy. We added 100 μg/ml UCN encapsulated and uncapsulated RE-1 peptide into HeLa cells for 24 h, then prepared samples and observed them by transmission electron microscopy. As shown in Fig. 5.12, in untreated cells, the cells were in good condition, the cell structure was normal, and there were very few vesicle structures in the cytoplasm; in 100 μg/ml UCN-treated HeLa cells, a large number of double-layer membrane-like autophagosome structures and autophagolysosome structures can be observed under transmission electron microscopy; while in the equivalent UCN-treated group encapsulated with RE-1 short peptide, the number

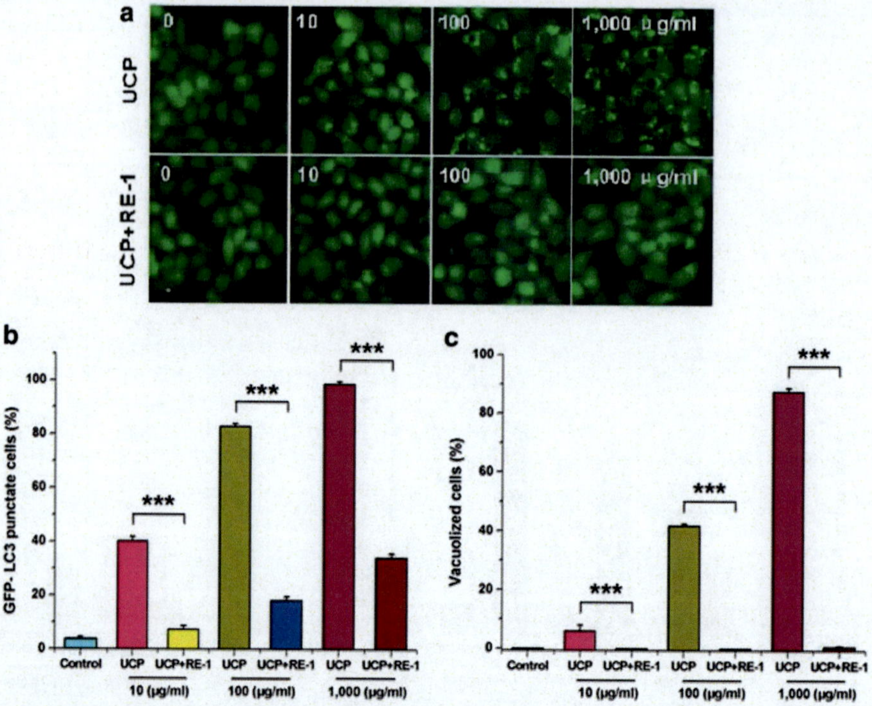

Fig. 5.10 RE-1 shields GFP-LC3/HeLa cells from punctate aggregation and cell vacuolization induced by UCP at different concentration gradients

of autophagosomes was greatly reduced, and the cells presented a normal structure similar to the control group cells. We replaced UCP, another kind of rare earth upconversion luminescent nanomaterial with larger particles, to repeat the above experiments, and the same results were obtained. As shown in Fig. 5.13, in HeLa cells treated with 100 μg/ml UCP, the cell state was poor, the edge of the nucleus was shrunken, and more obvious vesicle-like structures and transdermal vacuoles appeared in the cytoplasm. After enlargement, it was found that many of them were autophagosomes encapsulated with a double-layer membrane structure of nanoparticles, as well as a large number of cytoplasmic vacuoles. These vacuoles may be the vacuoles lost after mitochondrial swelling and degeneration; the intracytoplasmic structure of RE-1 short peptide in UCP-treated group was relatively normal, and vesicle structure was less. It further illustrates that RE-1 short peptide can inhibit the ability of rare earth upconversion luminescent nanomaterials to induce cell autophagy (autophagosome formation and cytoplasmic vacuolization effect).

Control was the untreated control cells; UCN was the UCN treatment group added with 100 μg/ml; UCN + RE-1 was the 100 μg/ml UCN treatment group encapsulated with 50 μg/ml short peptide. N represents the nucleus; arrows point to autophagosomes within cells. Scale: 1 um.

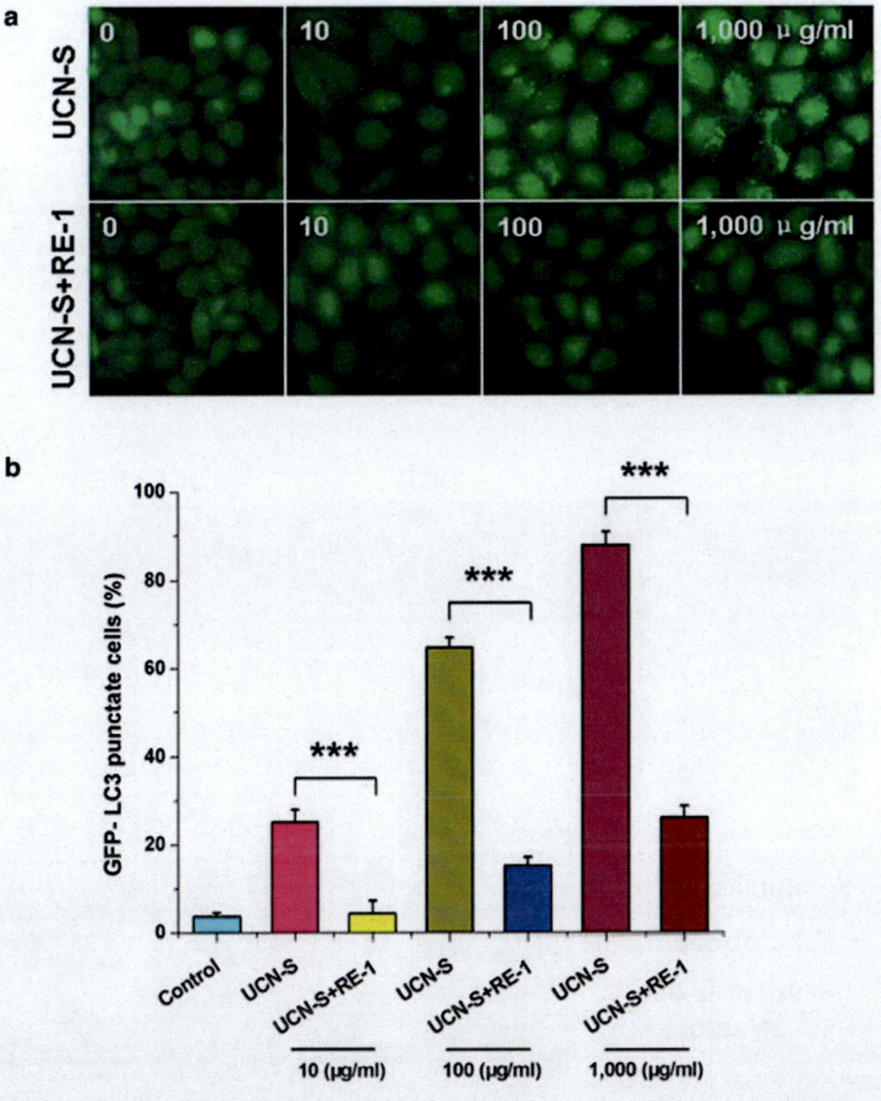

Fig. 5.11 RE-1 shields GFP-LC3 punctate aggregation and cell vacuolization in GFP-LC3/HeLa cells induced by UCN-S with different concentration gradients

Control was the untreated control cells; UCP was the UCP treatment group added with 100 μg/ml UCP; UCP + RE-1 was the UCP treatment group added with 100 μg/ml UCP encapsulated with 10 μg/ml RE-1 short peptide. Figures 5.1 and 5.2 are enlarged views of the black box area in the upper right corner and lower right corner of the UCP-treated group. N represents the nucleus; red arrows point

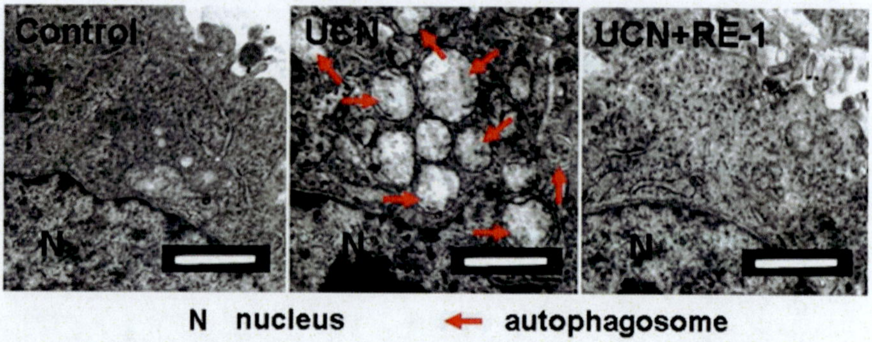

Fig. 5.12 Transmission electron microscopy showed that RE-1 short peptide inhibited UCN-induced autophagosome formation in cells

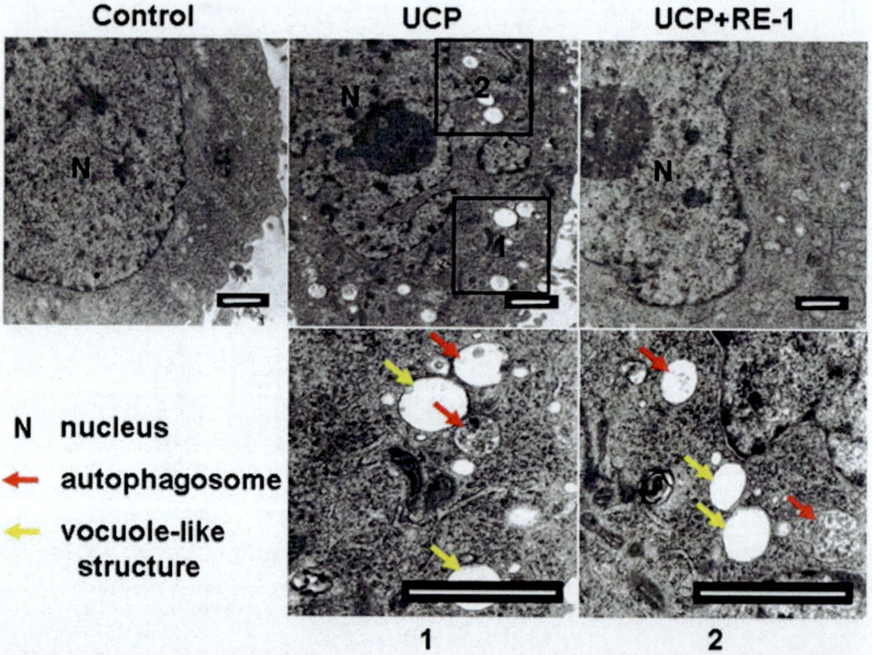

Fig. 5.13 Transmission electron microscopy showed that RE-1 short peptide inhibited UCP-induced autophagosome formation in cells

to intracellular autophagosomes; yellow arrows point to vacuoles in the cytoplasm. Scale: 1 μm.

To further prove this conclusion, we again applied Western Blot to compare the conversion between endogenous LC3-I and LC3-II and the expression level of LC3-II protein in HeLa cells treated with rare earth upconversion luminescent nanomaterials

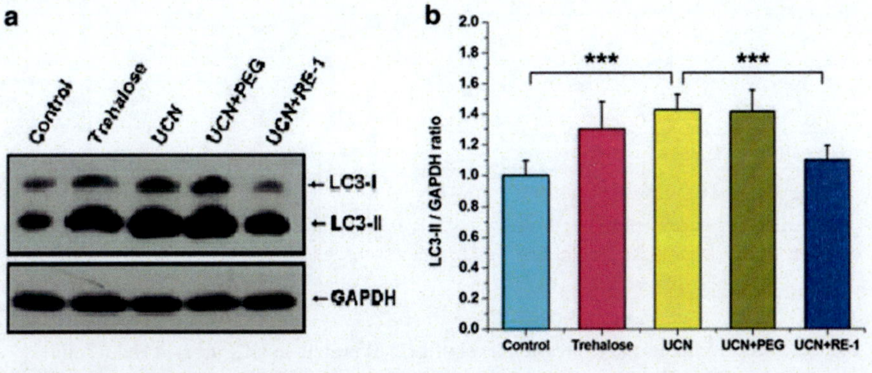

Fig. 5.14 RE-1 inhibits the expression level of LC3-II protein in UCN-treated HeLa cells

encapsulated with and without RE-1 short peptide. We added 100 μg/ml uncapsulated UCN and 50 μg/ml RE-1 short peptide encapsulated UCN into HeLa cells respectively, and collected the cells for Western Blot detection after 24 h of treatment. In order to obtain accurate and reliable experimental results, we set up a blank control group of untreated cells, a positive control group of 100 mM trehalose treatment, and a UCN negative control group of nonspecific adsorption of 10 mg/ml PEG with the same incubation method of RE-1.As shown in Fig. 5.14, UCN can activate the conversion of endogenous LC3-I to LC3-II in HeLa cells and increase the expression level of LC3-II protein; while RE-1 short peptide significantly attenuates the conversion of endogenous LC3-I to LC3-II in HeLa cells caused by UCN, and greatly reduces the expression level of LC3-II protein; as a PEG molecule is widely used to reduce the interaction between materials and organisms, Even its concentrations was up to 10 mg/ml, it did not significantly affect the conversion of endogenous LC3-I to LC3-II and the high expression level of LC3-II protein in UCN-induced cells. This result further illustrates that RE-1 short peptide can effectively inhibit UCN-induced cell autophagy ability.

Figure 5.14a is the result of Western Blot. The treatment order from left to right is: untreated control cells, 100 mM trehalose treatment group, 100 μg/ml UCN alone treatment group, 100 μg/ml UCN treatment group containing 10 mg/ml PEG, and 100 μg/ml UCN encapsulated with 50 μg/ml RE-1 treatment group; Fig. 5.14b is the quantitative statistical diagram of LC3-II/GAPDH in Fig. 5.14a, which can visually show the expression level of LC3-II protein.

At the same time, we also detected UCP, another rare earth upconversion luminescent nanomaterial slightly larger than UCN particles. To better determine whether the inhibitory effect of RE-1 on the ability of UCN to induce autophagy is polypeptide-specific and material-specific, we introduced two negative controls: material change, which replaced UCP with titanium dioxide (TiO_2), and polypeptide change, which replaced RE-1-specific binding peptide with AP-1 control peptide. 100 μg/ml RE-1 was mixed with 400 μg/ml TiO_2 by the same encapsulation method, and 10 μg/ml AP-1 and RE-1 were encapsulated on the surface of 100 μg/ml UCP, respectively,

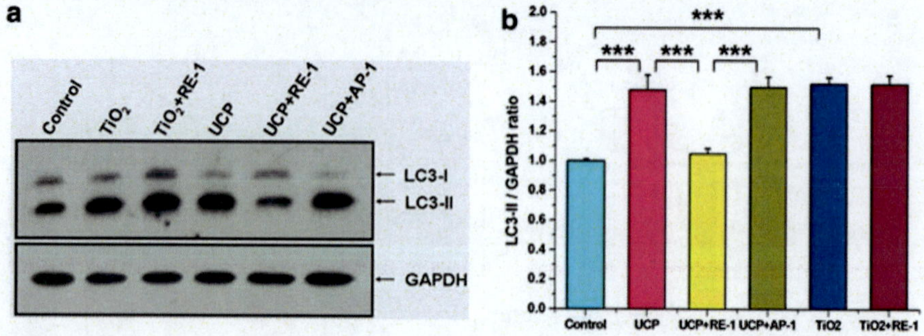

Fig. 5.15 RE-1 inhibits the expression level of LC3-II protein in UCP-treated HeLa cells

and added into HeLa cells. The untreated cells were set as negative control, and the 400 μg/ml TiO_2 alone treatment group was set as positive control. Samples were collected for Western Blot detection after 24 h of cell treatment. As shown in Fig. 5.15, consistent with the results obtained by UCN above, UCP activated the conversion of LC3-I to LC3-II, and the protein level of LC3-II was significantly increased; the same RE-1 short peptide significantly inhibited the UCP-induced conversion of LC3-I to LC3-II, greatly reduced the level of LC3-II protein, while the control peptide AP-1 had no such effect; as a control material, TiO_2 could also cause cell autophagy and activate the conversion of LC3-I to LC3-II, which was consistent with the previous report [3], but the RE-1 short peptide failed to inhibit the conversion of LC3-I to LC3-II initiated by TiO_2 and the ability to increase the level of LC3-II protein. This further demonstrates that the specific RE-1 short peptide has the ability to shield the rare earth upconversion luminescent nanomaterials from inducing cell autophagy, which is achieved by reducing the non-specific interaction between nanomaterials and cells. The RE-1 short peptide itself does not affect the whole autophagy pathway of the nanomaterials in inducing cell autophagy.

Figure 5.15a is the result of Western Blot. The treatment order from left to right is: untreated control cells, 400 μg/ml TiO_2 treatment group, 400 μg/ml TiO_2 treatment group with 100 μg/ml RE-1, 100 μg/ml UCP treatment group, 100 μg/ml UCP treatment group with 10 μg/ml RE-1, and 100 μg/ml UCP treatment group with 10 μg/ml AP-1. UCP treatment group; Fig. 5.15b is the quantitative statistical diagram of LC3-II/GAPDH in Fig. 5.15a, which can visually show the expression level of LC3-II protein.

5.5.4 *RE-1 Reduces the Toxicity of Rare Earth Upconversion Luminescent Nanomaterials in Cells*

It has been mentioned above that most of the autophagic effects induced by nanomaterials promote cell death, which deepens people's concerns about the safety of nanomaterials. In this paper, for the first time, we found that RE-1, a short peptide that specifically binds to RE upconversion luminescent nanomaterials, can effectively shield the autophagic effect induced by RE-1 upconversion luminescent nanomaterials in cells, so can RE-1 short peptide affect the toxic effect induced by nanomaterials in cells by reducing the autophagic ability of nanomaterials? To verify this hypothesis, cells treated with rare earth upconversion luminescent nanomaterials were tested by MTT cell viability assay and PI/Hoechst double staining to evaluate the cytotoxicity of rare earth upconversion luminescent nanomaterials. We added different concentration gradients of 10, 100, 250, 500, 1000 and 1000 μg/ml uncapsulated and encapsulated RE-1 peptide UCN into HeLa cells. After 24 h of treatment, MTT cell viability test and PI/Hoechst double staining were performed to detect cell death. As shown in Fig. 5.16, 250 μg/ml UCN-treated cells showed obvious cytotoxicity. When the concentration of UCN added into the cells was greater than 250 μg/ml to reach 1000 μg/ml, almost half of the cells were killed; after encapsulating RE-1, this toxic effect of UCN was completely eliminated. This indicates that RE-1 short peptide can effectively shield UCN-induced cytotoxicity. We further tested the UCP with slightly larger particles, and the results showed that UCP exhibited greater cytotoxicity than UCN. When the concentration of UCP was more than 100 μg/ml, the cells showed obvious cytotoxicity. When the concentration of UCP was added up to 1000 μg/ml, 70% of the cells died; and when UCP was encapsulated with RE-1, RE-1 also significantly inhibited the cytotoxicity of UCP (Fig. 5.17). The toxic effect of UCP over UCN shown in this result is consistent with that of another cytotoxic effect mentioned earlier that UCP can cause cytoplasmic vacuolization under TEM (Fig. 5.13), further illustrating that RE-1 can significantly reduce the cytotoxicity of rare earth upconversion luminescent nanomaterials. However, is this ability of RE-1 short peptides to reduce the cytotoxicity of rare earth upconversion luminescent nanomaterials achieved by shielding their induction? What is the correlation between this cytotoxicity and autophagy? To answer these questions, we tested the cytotoxicity exhibited by UCN under the action of the autophagy inhibitor 3-MA. As shown in Fig. 5.18, 3-MA could only inhibit 63% of the cytotoxicity caused by UCN, indicating that most of the cytotoxicity caused by UCN was caused by autophagic effect, while still some of the toxicity was not caused by autophagy induced by nanomaterials. It was demonstrated that UCN could simultaneously display dual toxicity depending on and independent of autophagy, both of which could be shielded by RE-1.

In conclusion, RE-1 can effectively reduce the toxic effects caused by rare earth upconversion luminescent nanomaterials in cells and improve its biological safety.

Figure 5.16a shows the quantitative statistics of MTT cell viability detection. Figure 5.16b shows the PI/Hoechst double-staining quantitative statistics.

Figure 5.17a shows the quantitative statistics of MTT cell viability detection. Figure 5.17b shows the PI/Hoechst double-staining quantitative statistics.

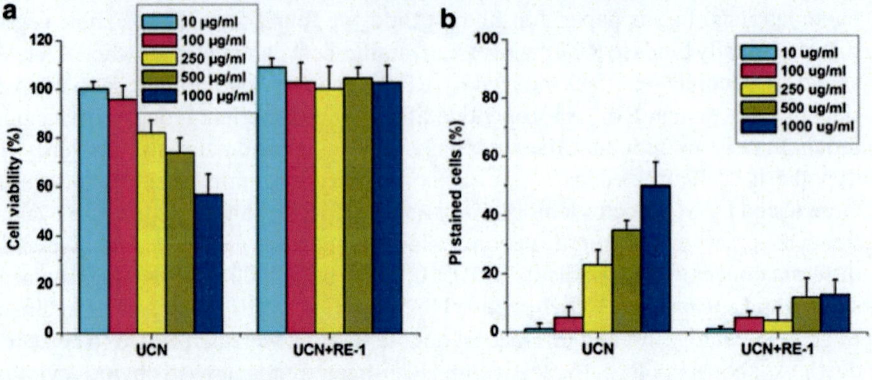

Fig. 5.16 RE-1 reduces UCN-induced cytotoxicity

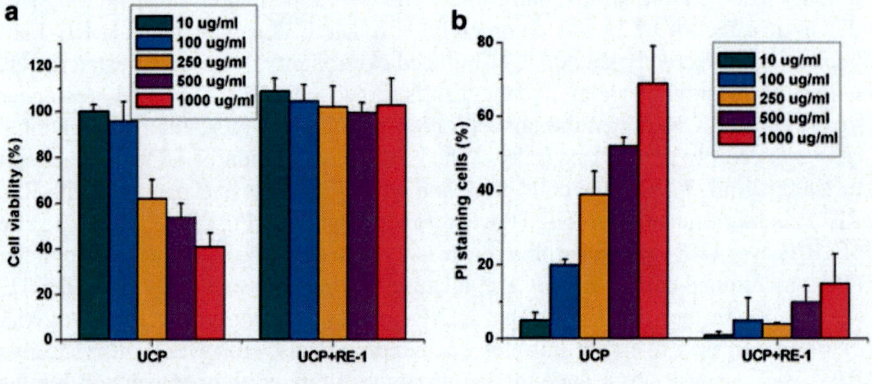

Fig. 5.17 RE-1 reduces UCP-induced cytotoxicity

Fig. 5.18 PI/Hoechst double staining quantitative statistical graph shows that 3-MA can reduce the cytotoxicity of UCN

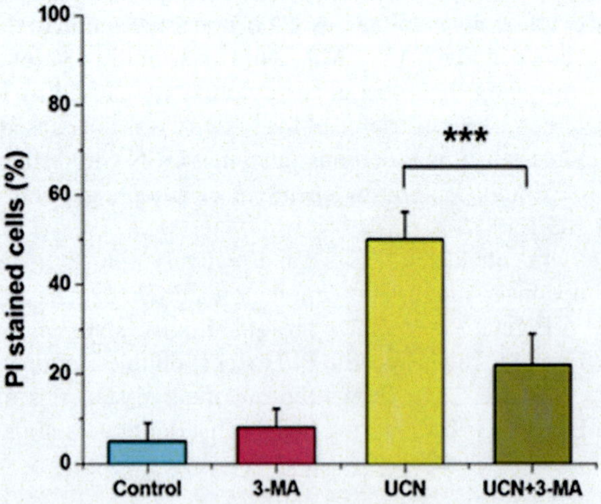

5.5.5 Effects of RE-1 Analogues on Autophagy and Cytotoxicity of Rare Earth Upconversion Luminescent Nanomaterials

To further determine the special ability of RE-1 shielding rare earth upconversion luminescent nanomaterials to induce autophagy and reduce cytotoxicity, we tested the effect of using RE-1 analogues on the autophagic capacity of UCN. RE-1 analogues RE-2, RE-3 and RE-5 were used to replace RE-1, and control group experiments were also set up: UCN was replaced by trehalose (Tre), an autophagy chemical inducer, and RE-1 was replaced by PEG. Here, to further demonstrate that the inhibition of RE-1 on the ability of UCN to induce autophagy is not due to the inhibition of the autophagy pathway by the RE-1 short peptide itself, we introduced a control group of trehalose and RE-1 short peptide mixtures. The results showed that similar peptides RE-2 and RE-5 did not inhibit the formation of GFP-LC3 punctate aggregation and cytoplasmic vacuolization of UCN in GFP-LC3/HeLa cells; PEG non-covalently adsorbed with the same encapsulation method of RE-1 had a weak inhibitory effect even when the concentration was up to 100 mg/ml, and the statistical results were not significantly different. ^{1}H NMR and FTIR experiments in Chap. 4 have proved that this encapsulation method can indeed make PEG adsorb to the surface of UCN nonspecifically, because the possibility that UCN does not affect autophagy due to PEG not connected to it is excluded here; after treatment of control Tre and RE-1 mixture group, no inhibition effect of RE-1 short peptide on Tre-induced cell autophagy was found, which indicates that RE-1 short peptide itself does not affect the process of autophagy; The peptide RE-3 showed some inhibitory effects on the formation of GFP-LC3 punctate aggregation and cytoplasmic vacuolization of UCN in GFP-LC3/HeLa cells. The statistical results showed that RE-3 could inhibit 21% of GFP-LC3 punctate aggregation of UCN in GFP-LC3/HeLa cells and 50% of cytoplasmic vacuolization (Fig. 5.19). This is consistent with the results of the relative activity of RE-3 in inhibiting the binding of REOB-1 phage to nano-Nd_2O_3 (as shown in Table 1) in Chap. 2. At the same time, we also tested the effect of RE-1 analogues on the ability of UCP to induce autophagy, and similar results were also obtained in inhibiting GFP-LC3 punctate aggregation and cytoplasmic vacuolization (Fig. 5.20). These results further confirm that RE-1 short peptides can effectively shield the formation of punctate aggregation and cytoplasmic vacuolization of rare earth upconversion luminescent nanomaterials in cells without affecting the autophagy process, and that RE-1 exhibits polypeptide specificity, which is not found in analogues RE-2, RE-5 and PEG.

Next, we evaluated the effects of RE-1 analogues and PEG on the cytotoxicity of rare earth upconversion luminescent nanomaterials. We added 500 mg/ml RE-1, RE-2, RE-3, RE-5 encapsulated with RE-1 and 1 mg/ml UCN non-covalently adsorbed 100 mg/ml PEG with the same encapsulation method of RE-1 into HeLa cells. After 24 h of treatment, the cell viability of treated cells was detected by MTT method. As shown in Fig. 5.21a, RE-1 significantly reduced the cytotoxicity caused by UCN, and RE-3 had a certain effect on inhibiting the cytotoxicity of UCN, while RE-2 and RE-5 were almost ineffective, and PEG also had no ability to reduce the cytotoxicity of UCN, consistent with the result of autophagy. For UCP with slightly larger particles, the effect of RE-1 analogues on its cytotoxicity was also examined. We obtained the same results as UCN. RE-1 had significant inhibitory activity on the cytotoxicity of UCP, RE-3 reduced the cytotoxicity of UCP to some extent, and RE-2 and RE-5 had no such effect (Fig. 5.21b). It is demonstrated that RE-1 has sequence specificity in reducing the cytotoxicity of rare earth upconversion luminescent nanomaterials.

Figure 5.19 Effect of coating with RE-1 variants on GFP-LC3 dot formation induced by UCN. Figure 5.19a shows that GFP-LC3/HeLa cells were treated for 24 h with trehalose (100 mM) plus RE-1 (1 mg/ml), uncoated UCN (1 mg/ml) or UCN (1 mg/ml) coated with PEG, RE-1 and the three variants. Figure 5.19b, c shows the quantified results.

Figure 5.20a is a fluorescence microscope photograph, in which Control represents untreated cells, UCP is 1 mg/ml UCP treatment group, and UCP + RE-1, UCP + RE-2, UCP + RE-3, UCP + RE-5 are 1 mg/ml UCP treatment group encapsulated with 100 mg/ml RE-1, RE-2, RE-3 and RE-5, respectively; Fig. 5.20b is a statistical

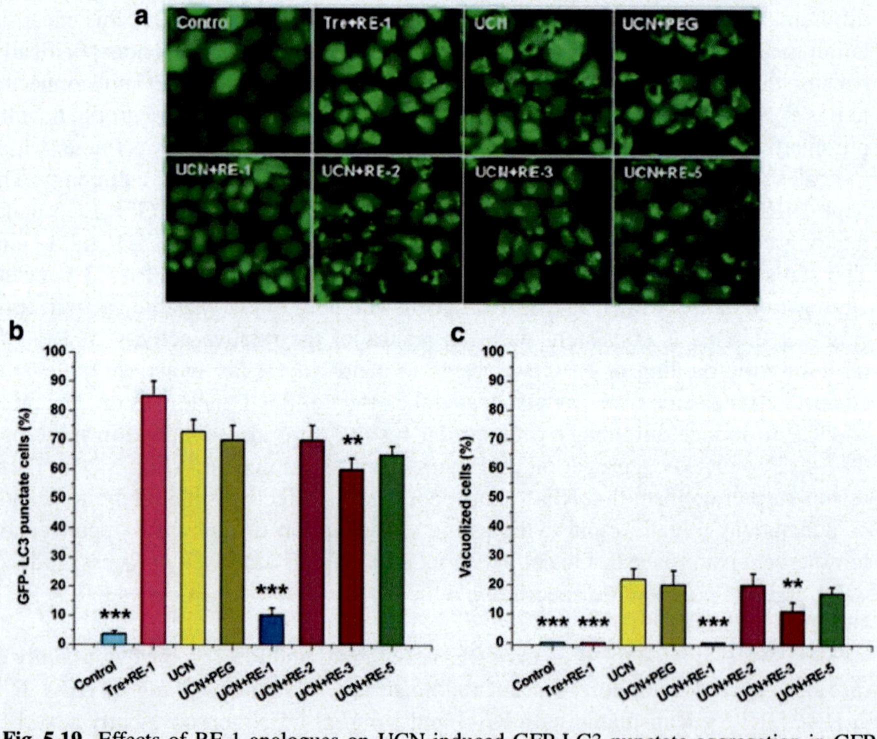

Fig. 5.19 Effects of RE-1 analogues on UCN-induced GFP-LC3 punctate aggregation in GFP-LC3/HeLa cells

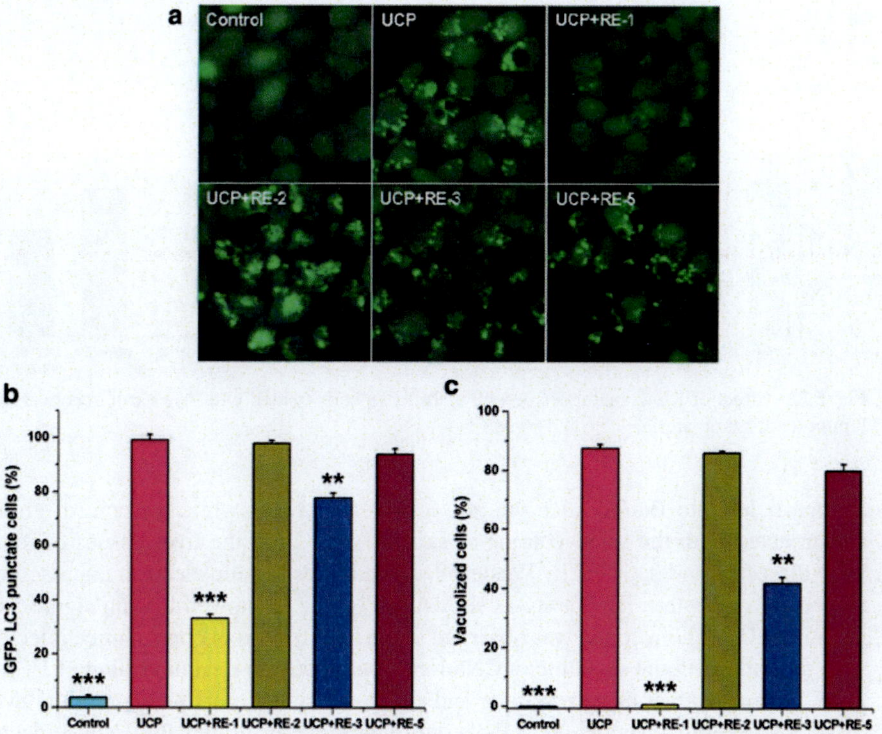

Fig. 5.20 Effects of RE-1 analogues on UCP-induced GFP-LC3 punctate aggregation in GFP-LC3/HeLa cells

map of GFP-LC3 dot aggregation in Fig. 5.20a; Fig. 5.20c is a statistical map of cytoplasmic vacuolation in Fig. 5.20a.

Figure 5.21a shows the quantitative statistics of UCN cell viability detection in different wrapping forms; Fig. 5.21b shows the quantitative statistics of UCP cell viability detection in different wrapping forms.

5.5.6 RE-1 Shielded Autophagy Induced by Rare Earth Upconversion Luminescent Nanomaterials in Liver Tissue of Mice

We have demonstrated that the specific binding short peptide RE-1 can effectively shield the autophagic effect induced by rare earth upconversion luminescent nanomaterials in HeLa cells. To further evaluate the special ability of RE-1, we examined its response in liver tissues of mice in vivo. We injected 15 mg/kg of uncapsulated and RE-1-encapsulated UCN nanoparticles and nonspecific PEG-adhering UCN

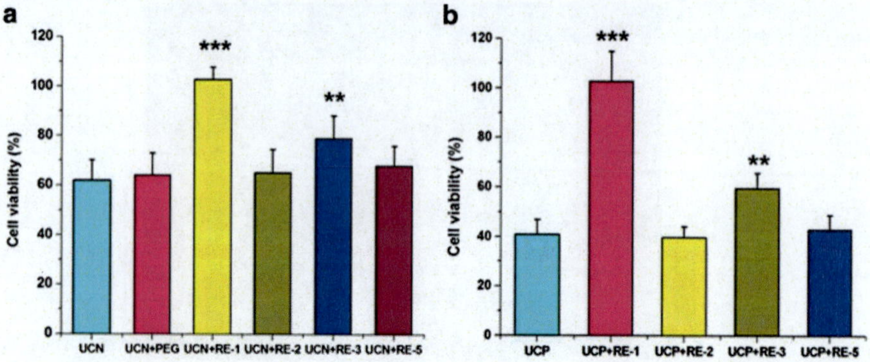

Fig. 5.21 Effect of RE-1 analogues on cell viability in cells treated with rare earth upconversion luminescent nanomaterials detected by MTT

nanoparticles into Bal/bc mice via tail vein, respectively, while the control group was injected with the same volume of saline. After 24 h, the liver tissues of mice were dissected and detected by Western Blot and transmission electron microscopy, respectively. Western Blot assay as shown in Fig. 5.22a showed that no significant LC3-I to LC3-II conversion was observed in the liver tissues of control mice injected with the same amount of saline; UCN alone could activate the conversion of LC3-I to LC3-II in the liver tissues of mice and significantly increase the expression level of LC3-II protein; meanwhile, RE-1 could significantly inhibit the UCN-induced conversion of LC3-I to LC3-II in the liver tissues of mice, and greatly reduced the expression level of LC3-II protein; PEG did not show such effect. As shown in the transmission electron microscopy results of Fig. 5.23a, in the liver tissues of saline control mice, the liver cells were in good condition, the cell structure was normal, and the mitochondria and other organelles with complete structure were clearly visible; the cytoplasm of the liver cells in the liver tissues of mice after UCN injection was loose and a large number of vesicular autophagosome structures could be observed, the autophagosome structure of these bilayer membranes. Similarly, it can be seen in the UCN-treated group encapsulated with PEG; while in the liver tissue of mice treated with UCN encapsulated with RE-1, the hepatocyte structure was intact, the cytoplasm was compact, and vesicle structure was scarcely seen. The above results demonstrated that UCN could activate the autophagic effect in mouse liver tissue, which was significantly inhibited by RE-1 short peptide.

We also examined the ability of UCP with slightly larger particles to induce autophagy in liver tissues in mice. In order to better evaluate the ability of RE-1 short peptide, we set up a UCP treatment group encapsulated with AP-1 control peptide, which was injected into Balb/c mice by tail vein at the same time as well as RE-1 uncapsulated and encapsulated UCP. After 24 h, the liver tissues of mice were dissected for Western Blot detection and transmission electron microscopy detection, respectively, and tissue paraffin sections were prepared. Analysis of Western Blot assay results showed that only very low conversion between LC3-I and LC3-II occurred in the liver tissues of saline control mice; UCP could activate the conversion of LC3-I to LC3-II in the liver tissues of mice and enhance the expression level of LC3-II protein; and after encapsulating RE-1, this conversion of LC3-I to LC3-II was inhibited, and the expression level of LC3-II protein was also significantly reduced. The AP-1 control peptide did not have this effect (Fig. 5.22b). Transmission electron microscopy showed that the liver tissues of mice 24 h after tail vein injection were in good condition and the cell structure was normal in the liver tissues of normal saline control mice. The cytoplasm of hepatocytes in the liver tissues of mice after UCP injection was loose, and a large number of obvious vesicle structures could be observed. After enlargement, the autophagosome structure of bilayer membranes could be seen, some containing the monolayer membrane structure of the contents may be the autophagolysosome after the fusion of autophagosome and lysosome, and a large number of damaged mitochondria (ridges showed broken and diffuse state) could also be observed; these structural features can also be seen in the UCP-treated group encapsulated with the control peptide AP-1; while in the liver tissue of mice treated with the UCP encapsulated with RE-1, the hepatocyte structure is intact and the cytoplasm is compact. The vesicle structure is very few, and the mitochondria with complete structure and good state can be clearly seen after enlargement (Fig. 5.23b). It was further demonstrated that RE-1 short peptide could effectively inhibit the autophagy effect in liver tissue of mice activated by rare earth upconversion luminescent nanomaterials, but neither PEG molecule nor AP-1 short peptide had this ability.

To further prove this special ability of RE-1, we also detected the endogenous LC3 protein expression level in mouse liver tissues by immunofluorescence. In the above experiments, paraffin sections prepared from liver tissues of Balb/c mice treated with AP-1 control peptide-encapsulated UCP, uncapsulated UCP and RE-1 short peptide-encapsulated UCP for 24 h were immunofluorescently stained, and endogenous LC3 protein was visualized under green light excitation by secondary antibody with red rhodamine, and observed by fluorescence microscopy. As shown in Fig. 5.24, a large amount of endogenous LC3 protein could be observed in mouse liver cells treated with UCP alone and UCP encapsulated with AP-1, while RE-1 could significantly inhibit the synthesis of endogenous LC3 protein activated by UCP in mouse liver cells (marking the formation of autophagosomes), that is, RE-1 could effectively shield the ability to induce autophagy in mouse liver cells.

Figure 5.22a shows the UCN treatment groups. The order of treatment from left to right was as follows: blank control group injected with equal amount of saline, UCN treatment group injected with 15 mg/kg UCN alone, UCN treatment group injected with 15 mg/kg UCN with non-specific adherent PEG, and UCN treatment group injected with 15 mg/kg UCN encapsulated RE-1 short peptide. Figure 5.22b shows the UCP-treated groups. The order of treatment from left to right was: blank control group injected with equal amount of saline, UCP treatment group injected with 15 mg/kg UCP alone, UCP treatment group injected with 15 mg/kg UCP encapsulated RE-1 short peptide, UCP treatment group injected with 15 mg/kg UCP encapsulated AP-1 control peptide. High exp. stands for strong exposure, low exp. stands for weak exposure.

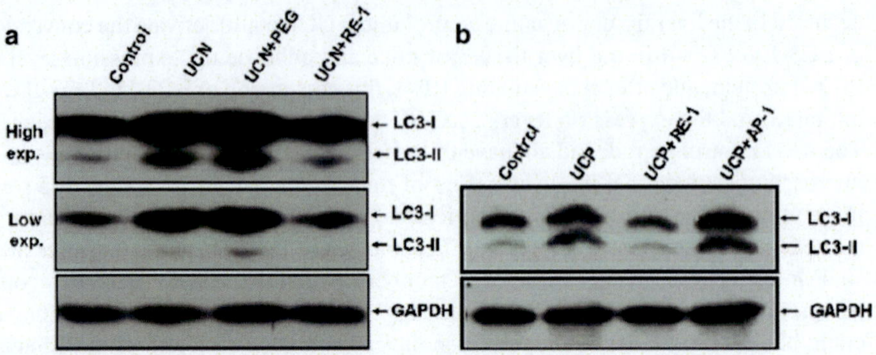

Fig. 5.22 Expression level of LC3-II protein in liver tissue of mice

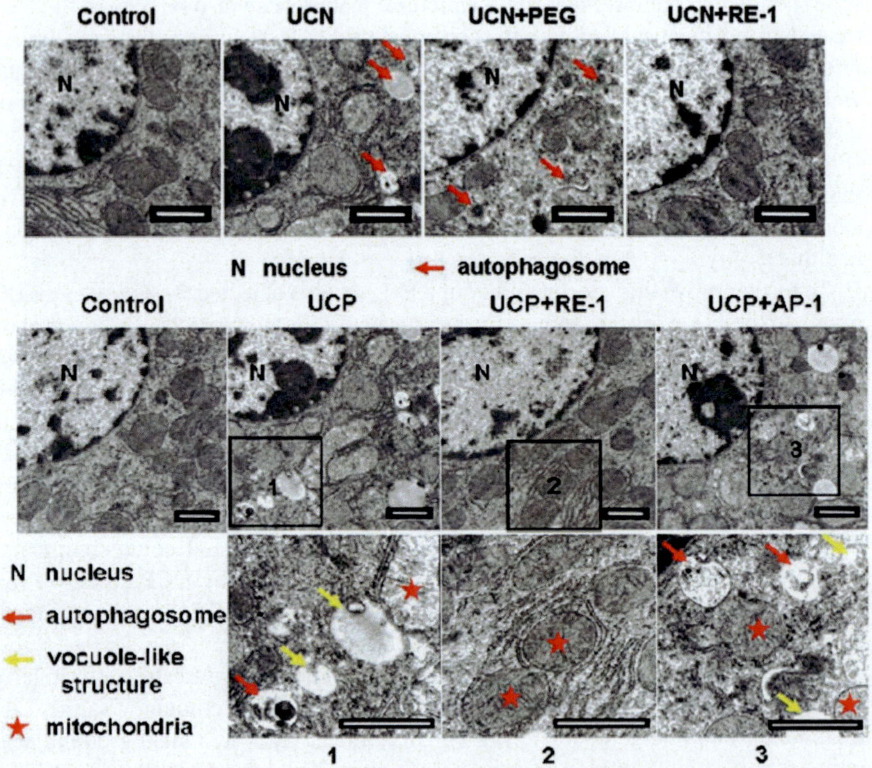

Fig. 5.23 Transmission electron microscopy of mouse liver tissue

Figure 5.23a shows the UCN treatment groups. Control was a blank control group injected with equal amount of saline, UCN was a separate treatment group injected with 15 mg/kg of UCN, UCN + PEG was a treatment group injected with 15 mg/kg UCN with nonspecific adherent PEG, and UCN + RE-1 was a treatment group injected with 15 mg/kg UCN with encapsulated RE-1 short peptide. Figure 5.23b shows the UCP-treated groups. Control was a blank control group injected with equal amount of saline, UCP was a treatment group injected with 15 mg/kg UCP alone, UCP + RE-1 was a treatment group injected with 15 mg/kg UCP encapsulated with RE-1 short peptide, and UCP + AP-1 was a UCP treatment group injected with 15 mg/kg encapsulated AP-1 control peptide. Figures 5.1, 5.2 and 5.3 in the lower row represent the enlarged areas in the black boxes in the upper figure, respectively. N represents the nucleus, red arrows point to autophagosomes, yellow arrows point to vacuolar structures, and asterisks represent mitochondria. Scale: 1 um.

Control was a blank control group injected with equal amount of saline, UCP was a treatment group injected with 15 mg/kg UCP alone, UCP + RE-1 was a treatment group injected with 15 mg/kg UCP encapsulated with RE-1 short peptide, and UCP + AP-1 was a treatment group injected with 15 mg/kg UCP encapsulated with AP-1 control peptide.

5.5.7 RE-1 Shields Liver Injury Induced by Rare Earth Upconversion Luminescent Nanomaterials in Mice

The above results of transmission electron microscopy of UCP-treated mouse liver tissue have shown that after the tail vein injection of UCP into mice for 24 h, a large number of damaged mitochondria in mouse liver tissue could be found, suggesting that rare earth upconversion luminescent nanomaterials may cause severe acute liver injury in mice. To prove this point, we performed HE (hematoxylin–eosin) staining on the above treated mouse liver tissues to evaluate the behavior of rare earth upconversion luminescent nanomaterials in mice from a pathological point of view. As shown in the pathological results of HE staining in Fig. 5.25, in the liver tissues of normal saline control mice, the liver cells were in good condition and the cell structure was normal, while in the liver tissues of mice after UCP injection, the liver injury was severe, the liver cells were deformed, the nuclei were pyknotic and multinucleated, and extranuclear multivacuolar degeneration was observed, accompanied by a large number of inflammatory cell infiltration. Similar structural features could also be observed in the UCP-treated group encapsulated with the control peptide AP-1; while in the UCP-treated group of mice encapsulated with RE-1, the hepatic cell structure was intact, the number of nuclei was normal, the cell state was basically good, and the degree of liver injury was significantly improved. Therefore, it can be demonstrated that UCP can cause severe acute liver injury in mice, and this liver injury can be inhibited by RE-1 short peptide, which further improves the biosafety

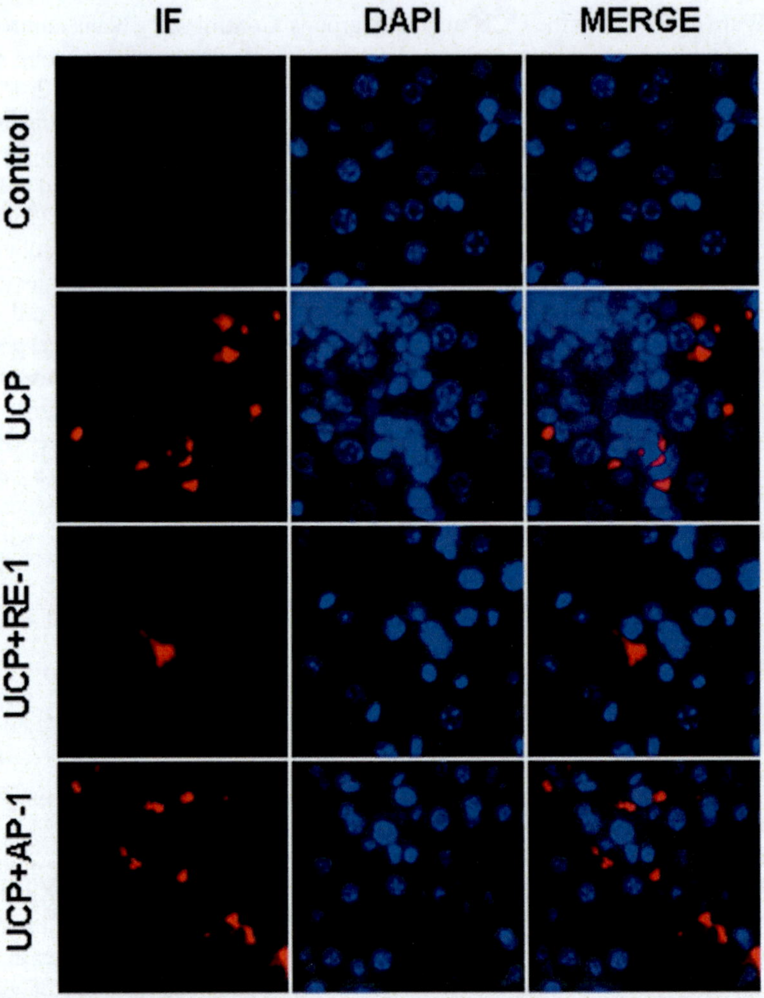

Fig. 5.24 Immunofluorescence (IF) shows the expression level of endogenous LC3 protein in mouse liver tissue

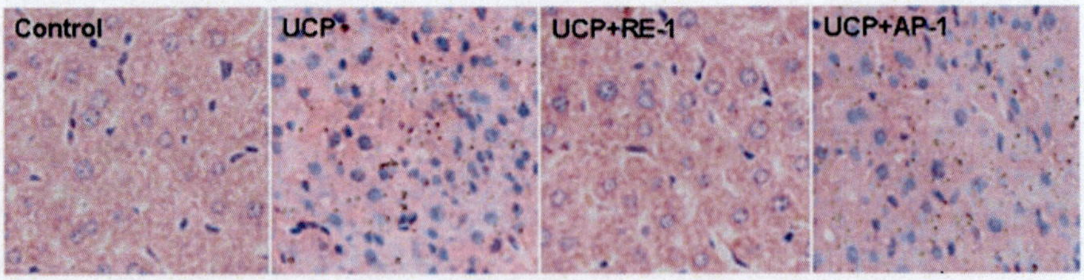

Fig. 5.25 Pathological results of HE staining of liver tissue injury in mice

of UCP in vivo application, and has great significance for the application of rare earth upconversion luminescent nanomaterials in vivo.

Control was a blank control group injected with equal amount of saline, UCP was a treatment group injected with 15 mg/kg UCP alone, UCP + RE-1 was a treatment group injected with 15 mg/kg UCP encapsulated RE-1 short peptide, and UCP + AP-1 was a treatment group injected with 15 mg/kg UCP encapsulated AP-1 control peptide.

5.6 Summary

In this chapter, the ability of rare earth upconversion luminescent nanomaterials to induce autophagy in vivo and in vitro was systematically studied. Firstly, the green fluorescent punctate aggregation of GFP-LC3 in GFP-LC3/HeLa cells induced by rare earth nanomaterials (UCN, UCP, and UCN-S) was observed by static detection method and the formation of autophagosomes was observed by transmission electron microscopy, indicating that rare earth upconversion luminescent nanomaterials can indeed induce the occurrence of autophagy. Secondly, through the treatment of autophagy inhibitor 3-MA and lysosome fusion inhibitor Bafilomycin A1, respectively, the autophagy induced by rare earth upconversion luminescent nanomaterials was detected by dynamic detection methods of autophagy such as detection of degradation substrate P62 and co-localization of autophagy-related components. It was found that the autophagy induced by rare earth upconversion luminescent nanomaterials was a complete process. However, the autophagic ability induced by this rare earth upconversion luminescent nanomaterial can be effectively shielded by encapsulating RE-1 short peptides. Through the detection of endocytosis ability of rare earth upconversion luminescent nanomaterials, we found that the ability of specific RE-1 short peptide shielding the materials to induce cell autophagy was achieved by reducing the non-specific interaction between nanomaterials and cells, and the RE-1 short peptide itself did not affect the whole autophagy pathway of nanomaterials to induce cell autophagy.

As mentioned earlier, autophagy is a "double-edged sword". When cells are stimulated by the external environment, cell autophagy can induce cell damage, leading to cell death. And most of the cell autophagy induced by nanomaterials is pro-cell death, which deepens people's concern about the safety of nanomaterials. Therefore, in this chapter, we further evaluate the toxic effects caused by rare earth upconversion luminescent nanomaterials. We found that rare earth upconversion luminescent nanomaterials exhibited obvious cytotoxicity at medium and high concentrations, but the cytotoxicity were all shielded by the encapsulated short peptide RE-1. We also found that RE-1 could shield both autophagy-dependent and autophagy-independent cytotoxicity caused by the materials.

In this chapter, we also investigated the ability of rare earth upconversion luminescent nanomaterials to induce autophagy in liver tissues of Balb/c mice in vivo. We found that both UCN and UCP could activate the autophagic effect in mouse liver tissues, which was also significantly inhibited by the short peptide RE-1. Meanwhile, the short peptide RE-1 can also shield the acute liver injury induced by rare earth upconversion luminescent nanomaterials in mice.

In summary, RE-1 has been demonstrated in this chapter to effectively shield the cell autophagy of nanomaterials and the resulting toxic effects, thus improving the biosafety of nanomaterials. Therefore, the biosafety problem of nanomaterials can be solved by artificial regulation of autophagy effect. At the same time, a systematic and in-depth study of the effect and mechanism of nanomaterials in regulating autophagy and its relationship with biosafety, and the establishment of corresponding detection systems and evaluation criteria are of great significance for the establishment of a biosafety evaluation system for nanomaterials to guide and standardize the production and safe application of nanomaterials.

References

1. Zhang, Y., et al. (2010). Nano rare-earth oxides induced size-dependent vacuolization: An independent pathway from autophagy. *International Journal Nanomedicine, 5*, 601–609.
2. Ma, X., et al. (2011). Gold nanoparticles induce autophagosome accumulation through size-dependent nanoparticle uptake and lysosome impairment. *ACS Nano, 5*(11), 8629–8639.
3. Yu, J. X., & Li, T. H. (2011). Distinct biological effects of different nanoparticles commonly used in cosmetics and medicine coatings. *Cell Bioscience, 1*(1), 19–27.

Chapter 6
RGD-RE-1 Bifunctional Short Peptide Enhances the Interaction Between Rare Earth Nanomaterials and Cancer Cells and the Effect of Cell Autophagy

6.1 Introduction

In the previous chapter, we employed a large number of experiments to prove that the specific binding short peptide RE-1 can effectively shield the ability of materials to induce autophagy in cells and mice by reducing the non-specific interaction between rare earth nanomaterials and cells, as well as rare earth nanomaterials, and improve the biological safety of materials. However, with the increasing application of rare earth nanomaterials in the biomedical field and the urgent need for the treatment of major diseases such as cancer, the development of nanomaterials that can enhance specific interactions and further specifically kill tumor cells is of great significance for the diagnosis and treatment of major diseases such as cancer. RGD peptides are a class of short peptides composed of arginine-glycine-aspartate (Arg-Gly-Asp), which can serve as recognition sites for the interaction of integrins (and their ligands) specifically expressed by tumor cells or neovasculature, mediate the sacrificial treatment of tumors, and are widely used in the modification of antitumor drugs and related drug carriers. In this chapter, we will describe in detail the effects of bifunctional complex short peptide RE-1-RGD synthesized by cancer cell targeting peptide RGD and short peptide RE-1 on the autophagy and toxicity of rare earth upconversion luminescent nanomaterials in integrin-highly expressed cells, which is expected to achieve the goals of shielding autophagy in normal cells and improving autophagy in target cells simultaneously by targeting strategy, thus providing theoretical basis for rare earth nanomaterials in cancer treatment.

6.2 Experimental Materials

6.2.1 Reagent

LC3 antibody (NB100-2220, Novus, USA)

Y. Zhang, *Tuning Autophagy-Inducing Activity and Toxicity for Lanthanide Nanocrystals*, Springer Theses, https://doi.org/10.1007/978-981-16-8166-0_6

GAPDH antibody (ab8245, Abcam, USA)
ECL development Kit (EZ-ECL, Biological Industries, Israel)
Protein Marker (94KD), cell lysate and powerful tissue lysate were all purchased from China Beyotime Biotechnology Research Institute.
Medium powder, serum and trypsin used for cell culture were purchased from Invitrogen Company.

6.2.2 Cell Line

HeLa cells: human cervical cancer cell line (purchased from Shanghai Institute of Biochemistry, Chinese Academy of Sciences)

GFP-LC3/HeLa cells: cell lines stably transfected with eGFP-LC3 plasmid, in which eGFP-LC3 plasmid was donated by Professor Noboru Mizushima, Tokyo Metropolitan Institute of Clinical Medicine, Japan.

6.2.3 Experimental Device and Consumables

Ultraviolet detection cell (X72053, ALPHA, UK)
Nitrocellulose membrane (Millipore, USA)
Cell culture dishes and cell culture plates related to cell experiments were purchased from Corning Company, USA.

6.3 Instruments and Equipment

Ultraviolet spectrophotometer (DU-640, Beckman, USA)
980 nm infrared laser (MDL-980 nm, 1 W, Changchun New Industries Optoelectronics Technology Co., Ltd., China)
Fluorescence microscopy (Olympus IX71, Olympus, Japan)
Vortex mixer (MS1 Minishaker, IKA, China)
Microcentrifuge (Centrifuge 5415D, Eppendorf, Germany)
Transmission electron microscopy (JEOL-1230, JEOL, Japan)
Electrophoresis apparatus (Tanon EPS300, TANON, China)

6.4 Experimental Methods

Please refer to other chapters for the same experimental method.

6.4.1 Combination of RE-1-RGD with UCN

After adding 50 μg RE-1-RGD or fluorescent peptide Rho-RE-1-RGD into 100 μL UCN (1 mg/ml) and mixing evenly, gently shake them at room temperature and incubate for 1 h. After centrifugation at 12,000 rpm for 10 min, the precipitated nanoparticles were washed 2–3 times with Milli-Q water and then resuspended with appropriate water or solvent according to the type of experiment for different experiments.

6.4.2 Establishment of a Standard Curve for the Relationship Between RE-1-RGD Fluorescence Value and Concentration

Rho-RE-1-RGD polypeptide was diluted into six different concentration gradients, and the peak values at 590 nm of each gradient sample were detected by 530 nm excitation spectroscopy in a fluorescence spectrophotometer, and the standard curve was drawn by the relationship between Rho-RE-1-RGD fluorescence value and concentration. The relationship equation between fluorescence value x and concentration y was obtained by fitting standard curve as follows: $y = 0.0009x + 0.0003$ ($R^2 = 0.9994$).

6.4.3 Identification Method of Binding Concentration of RE-1-RGD to UCN

100 μG UCN was mixed with different concentrations (5, 10, 25, 50 and 100 μg) of Rho-RE-1-RGD and incubated with gentle shaking at room temperature for 1 h. After centrifugation at 12,000 rpm for 10 min, the precipitated nanoparticles were washed 2–3 times with Milli-Q water, and the excess unbound peptides were washed away. Finally, they were resuspended in 100 μl of water, gently sonicated for 1 min, and then Rhodamine was detected by fluorescence spectrophotometer (excitation wavelength was 530 nm, emission wavelength was 590 nm). The standard curve equation of the relationship between the fluorescence value of Rho-RE-1-RGD and the concentration was used to obtain the amount of Rho-RE-1-RGD fluorescent peptide on UCN surface.

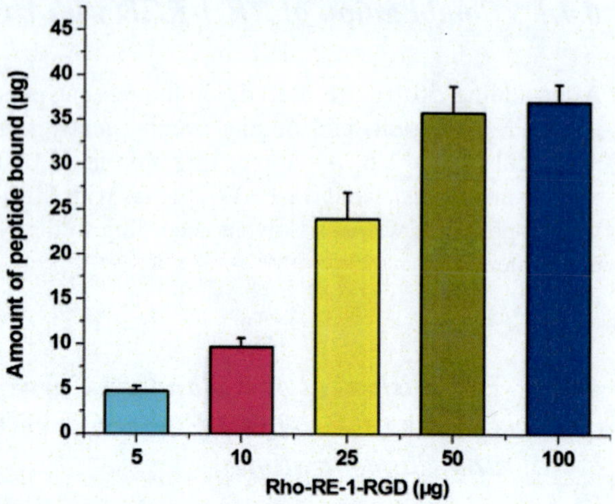

Fig. 6.1 Dose effect of RE-1-RGD binding to UCN

6.5 Experimental Results and Discussion

6.5.1 Determination of the Binding Concentration of RE-1-RGD to UCN

Through the study of method 6.4.3, we determined the binding concentration of RE-1-RGD to UCN. As shown in Fig. 6.1, the experimental results of this dose effect were further analyzed and calculated, and we found that the amount of 100 μg UCN capable of binding Rho-RE-1-RGD short peptide was 37 μg (about 16 nmol). Compared with 30 nmol, the amount of 100 μg UCN binding FITC-RE-1 peptide in Chap. 3, the amount of RE-1-RGD binding on UCN surface is only half of that of RE-1, suggesting that the three amino acids of RGD we linked reduce the binding ability of RE-1-RGD to UCN.

6.5.2 RE-1-RGD, like RE-1, Can also Reduce the Sedimentation Rate of UCN

It was found in the binding experiments above that the three amino acids of RGD linked reduced the binding ability of RE-1-RGD to UCN. So, will the three amino acids linked affect the suspension and sedimentation capacity of UCN in water? We further examined the effect of RE-1-RGD on the suspension capacity of UCN in water using time-kinetic curves. The absorption of 500 nm was detected by ultraviolet spectrophotometer for UCN without and with RE-1-RGD and UCN with RE-1, respectively, and the curves were continuously scanned for 120 min. As shown in Fig. 6.2, by analyzing the detection results of time kinetics, we found that, consistent

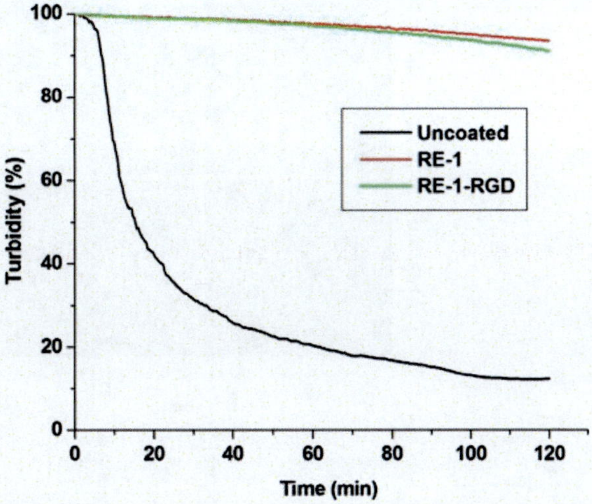

Fig. 6.2 Time-kinetics curve of RE-1-RGD reducing UCN sedimentation rate

with RE-1 short peptide, RE-1-RGD could also effectively reduce the sedimentation rate of UCN and enhance the suspension ability of UCN.

6.5.3 *RE-1-RGD Enhances the Specific Interaction Between UCN and Cells*

Since RGD can specifically interact with extracellular integrins, we first evaluated the effect of RE-1-RGD bifunctional short peptides on the interaction between UCN and cells. In order to obtain more accurate experimental results, we used the device used in Chap. 4 to detect the interaction of rare earth nanomaterials with cells through sedimentation and diffusion (Fig. 4.2), and compared the interaction of UCN encapsulated with RE-1-RGD and UCN encapsulated with RE-1 with cells under different conditions, cell-up or down. As shown in Fig. 6.3, unlike RE-1, the RE-1-RGD bifunctional peptide greatly enhances the interaction between UCN and cells. At the same time, we found that after encapsulating RE-1-RGD, the amount of UCN bound on the cell surface was essentially equal in both cell-up and cell-down cases, which indicated that UCN encapsulated RE-1-RGD had no sedimentation ability. UCN, which reached the cell surface by diffusion, enhanced the specific adhesion to the cell through the specific binding of RGD and extracellular integrin. To further prove this conclusion, we further designed identification experiments. We first pre-incubated RE-1-RGD at different concentrations (0, 10, 50, 100 and 500 μg/ml) with cells for 2 h before adding samples, then added UCN encapsulated with RE-1-RGD into cells, and then detected the amount of UCN adhering to the cell surface. As shown in Fig. 6.4, when the pre-incubated concentration of RE-1-RGD was higher than 100 μg/ml, the same level of shielding UCN from cell-to-cell interaction with

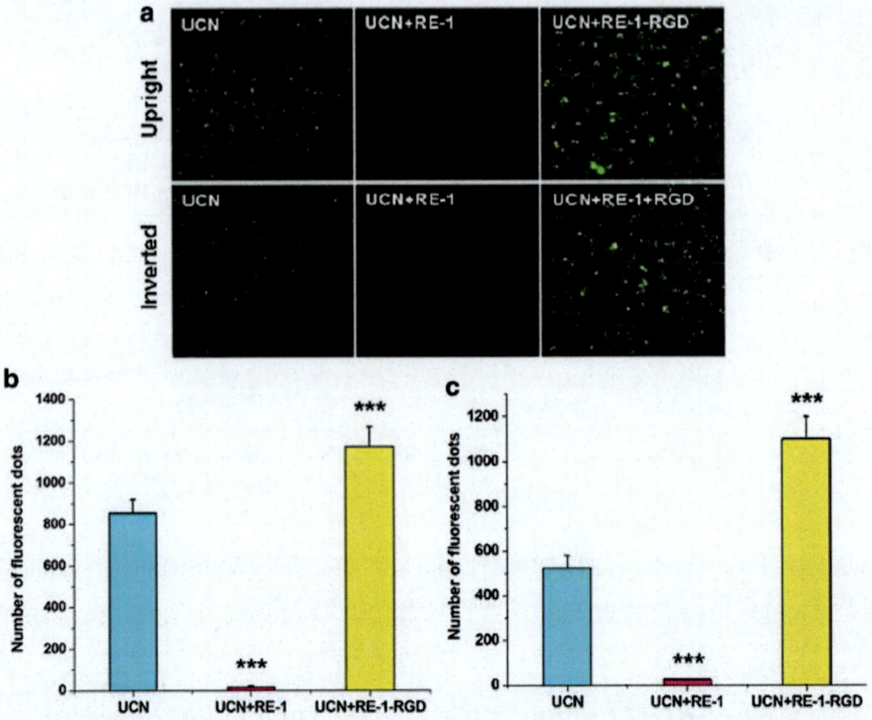

Fig. 6.3 Fluorescence microscopic pavilion identification of specific interaction of UCN encapsulating RE-1-RGD with up and down cells

Fig. 6.4 Effect of pre-incubated RE-1-RGD on the interaction between UCN and cells encapsulated with RE-1-RGD

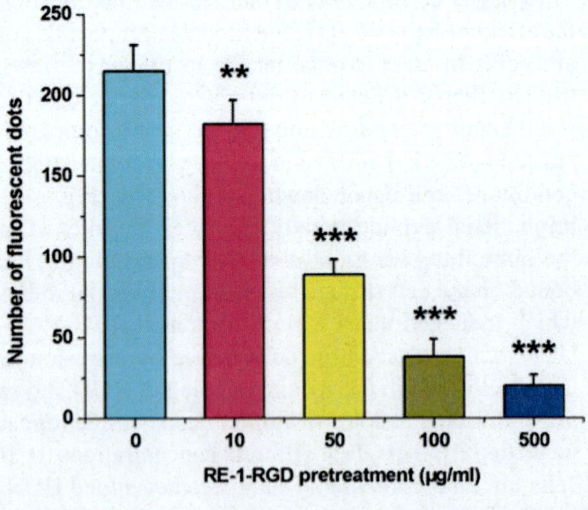

RE-1 could be achieved. This result intuitively illustrates that excessive RE-1-RGD bifunctional peptide can supersaturate the binding of RGD to integrins on the cell surface, thus reducing the interaction force between UCN and cells encapsulating RE-1-RGD, thereby reducing the uptake of UCN by cells. It was further demonstrated that the high specific adhesion ability of UCN encapsulating RE-1-RGD to cells was caused by the ligation of the integrin-specific binding peptide RGD, which also demonstrated that the promoting effect of the interaction force between UCN and cells caused by the RE-1-RGD bifunctional peptide could be artificially regulated by the RE-1-RGD peptide itself.

Figure 6.3a is a fluorescence microscopy image to directly observe the green fluorescence of UCN, Fig. 6.3b is a quantitative statistical map of UCP green inflammation spot when the cell is up, and Fig. 6.3c is a quantitative statistical map of UCP green fluorescence spot when the cell is down. Magnification factor is *100.

6.5.4 *RE-1-RGD Enhances the Ability of UCN to Induce Cell Autophagy and Its Cytotoxicity*

The studies in the previous chapters in this paper have shown that the specific binding short peptide RE-1 effectively shields the autophagic effect induced by rare earth nanomaterials in cells and mice by reducing the non-specific interaction between rare earth nanomaterials themselves and between rare earth nanomaterials and cells, and further reduces the toxicity of rare earth nanomaterials in cells and animals. So, for the RE-1-RGD bifunctional peptide that can enhance the interaction between UCN and cells, can it further enhance the ability of UCN to induce autophagy in cells with high integrin expression? We all know that most of the cells with high expression of integrin are tumor cells, if the above inference holds, then this experiment will provide a significant opportunity for rare earth nanomaterials in cancer diagnosis and treatment. To prove this point, we further evaluated the ability of UCN to induce cell autophagy after encapsulating RE-1-RGD. First, we examined the ability of UCN to induce cell autophagy after encapsulation of RE-1-RGD by observing the punctate aggregation of GFP-LC3. As shown in Fig. 6.5, we found that the punctate aggregation of GFP-LC3 and vacuolation in the cytoplasm of UCN-treated cells encapsulated with RE-1-RGD were significantly higher than those of UCN-treated cells alone, and much higher than those of UCN-treated cells encapsulated with RE-1. This result is consistent with our hypothesis, indicating that RE-1-RGD significantly improves the ability of UCN to induce cell autophagy. To further confirm this conclusion, we also used Western Blot and assayed the conversion of LC3-I to LC3-II and the expression level of LC3-II protein. The results showed that UCN encapsulated RE-1-RGD activated the conversion of LC3-I to LC3-II in HeLa cells, and RE-1-RGD greatly increased the expression level of medium LC3-II protein in UCN-treated HeLa cells, which strongly demonstrated that RE-1-RGD enhanced the ability of UCN to induce cell autophagy. In order to achieve the therapeutic

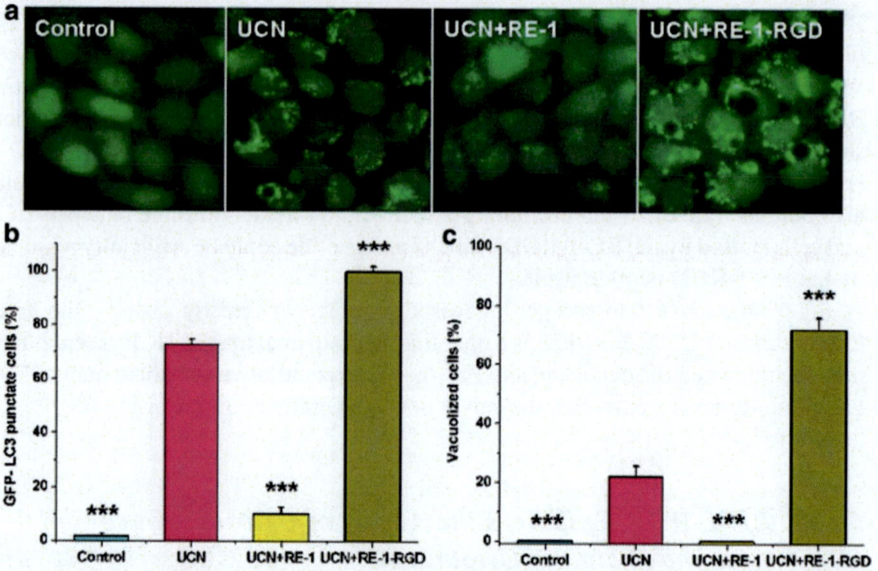

Fig. 6.5 GFP-LC3 punctate aggregation and cytoplasmic vacuolization of UCN in GFP-LC3/HeLa cells after encapsulation of RE-1-RGD

purpose of killing tumor cells by UCN encapsulating RE-1-RGD, we also tested the effect of RE-1-RGD on UCN-induced cytotoxicity. Through MTT cell viability detection, we found that RE-1-RGD doubled the cytotoxicity of UCN in HeLa cells (Fig. 6.7), and successfully achieved specific killing of tumor cells, which provided a significant opportunity for rare earth nanomaterials in cancer diagnosis, cancer treatment, and improving cancer efficacy. Finally, we evaluated the effect of pretreatment of cells with RE-1-RGD on the ability of UCN encapsulating RE-1-RGD to induce autophagy. As shown in Fig. 6.8, the punctate aggregation of GFP-LC3 and the vacuolation in cytoplasm induced by UCN encapsulating RE-1-RGD in GFP-LC3/HeLa cells could be attenuated with the increasing concentration of pre-treated RE-1-RGD bifunctional peptide. When the concentration of RE-1-RGD pre-incubated with cells reached 100 μg/ml, it could inhibit 80% of the punctate aggregation (the number of autophagosomes formed), and could all inhibit the vacuolation in cytoplasm. It is further illustrated that the ability of RE-1-RGD bifunctional peptide to promote UCN-induced autophagy and kill tumor cells can also be artificially regulated by RE-1-RGD peptide itself.

Figure 6.5a shows that Gfp-lc3 aggregation and cytoplasmic vacuolation of UCN coated with RE-1-RGD in GFP-LC3/HeLa cells. Figure 6.5b, c shows the quantified results.

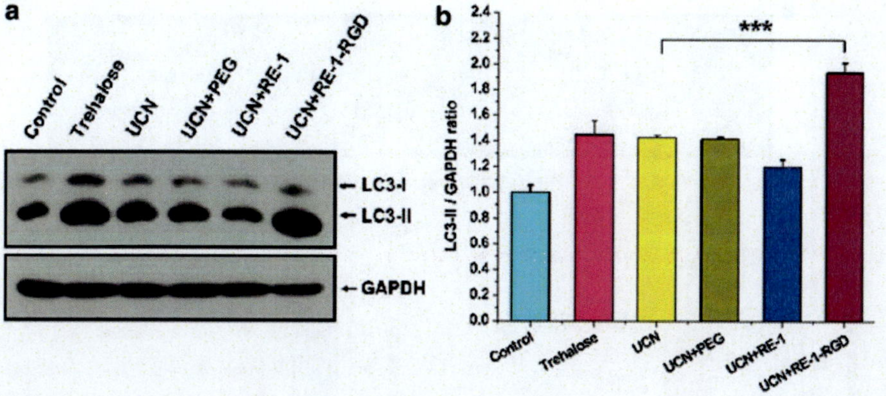

Fig. 6.6 RE-1-RGD increased the expression level of LCE-II protein in UCN-treated HeLa cells

Figure 6.6a is the result of Western Blot. The treatment order from left to right is: untreated control cells, 100 mM trehalose treatment group, 100 μg/ml UCN treatment group alone, 100 μg/ml UCN treatment group containing 10 mg/ml PEG, 100 μg/ml UCN encapsulating 50 ug/ml RE-1 treatment group, and 100 μg/ml UCN encapsulating 50 μg/ml RE-1-RGD treatment group; Fig. 6.6b is the quantitative statistical diagram of LC3-II/GAPDH in Fig. 6.6a, which can visually show the expression level of LC3-II protein.

Fig. 6.7 RE-1-RGD enhances UCN-induced cytotoxicity

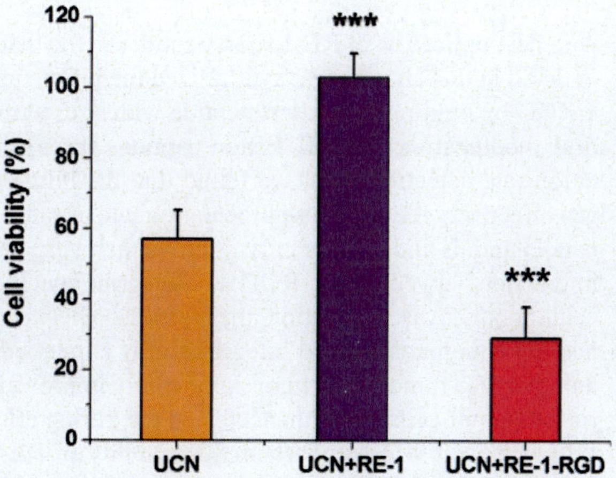

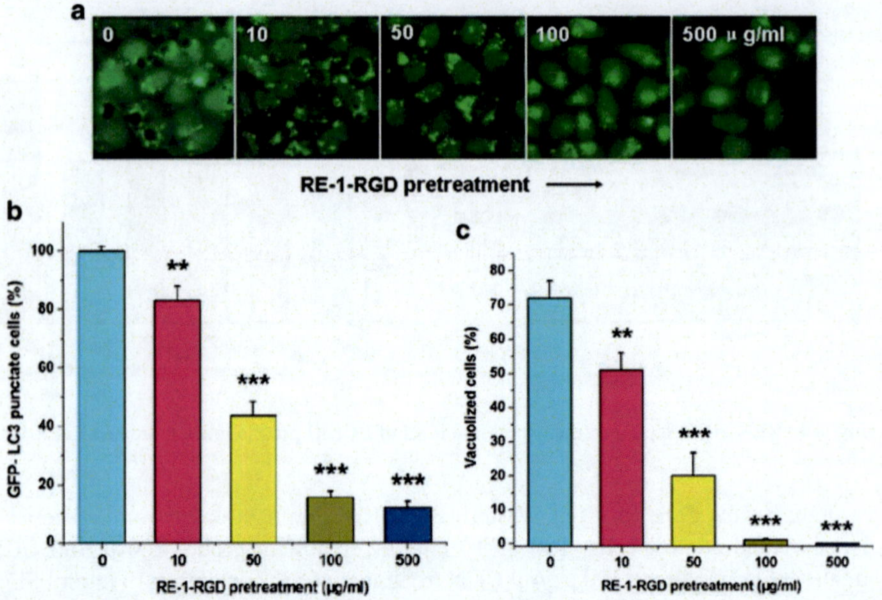

Fig. 6.8 Effect of pre-incubated RE-1-RGD on the ability of UCN encapsulated with RE-1-RGD to induce autophagy in GFP-LC3/HeLa cells

6.6 Summary

Another application of RE-1 short peptides in the field of cancer therapy is described in detail in this chapter. We synthesized the bifunctional complex short peptide RE-1-RGD by linking RGD short peptide widely used in tumor therapy and anti-tumor drug modification with RE-1 short peptide. Through the time kinetics and sedimentation rate detection tests, we found that the bifunctional composite short peptide can effectively reduce the sedimentation rate of rare earth upconversion luminescent nanomaterials and significantly improve their suspension performance. Through the interaction between RE-1-RGD and integrins specifically expressed in tumor extracellular or neovascularization, the interaction between rare earth upconversion luminescent nanomaterials and integrin-highly expressed cells and the ability to induce autophagy in tumor cells were significantly improved, the cytotoxicity of nanomaterials to tumor cells was enhanced, and the killing effect on tumor cells was successfully achieved. Most importantly, the ability of RE-1-RGD to enhance autophagy and kill cells of nanomaterials is artificially regulated by the RE-1-RGD peptide itself. It shows that the targets of shielding autophagy in normal cells and improving autophagy in target cells can be achieved simultaneously by targeting strategy, which provides significant opportunities for rare earth nanomaterials in cancer diagnosis and treatment and improves the therapeutic effect of nanomaterials on tumors.

Chapter 7
Summary and Outlook

7.1 Summary

In this article, we demonstrated the feasibility of using phage display technology for surface engineering of nanomaterials. The short peptide RE-1 that specifically binds to rare earth metal upconversion luminescent materials with high affinity was successfully screened, and it was proved that the combination of RE-1 short peptide and rare earth nanomaterials has both sequence specificity and material specificity. Through a series of analysis such as electron microscope observation, we found that the short peptide can form a coating on the surface of nanocrystals, which can reduce the interaction between nanomaterials and between nanomaterials and cells. Meanwhile, the nanomaterials can reduce the sedimentation and sedimentation of rare earth nanomaterials in water, and improve suspension capacity. Therefore, the autophagy of nanomaterials and the toxic effects caused by the nanomaterials can be shielded.

The modification of RE-1 polypeptide shows significant advantages over other surface-modified polymers (such as PEG, etc.) that are widely used to alter cell–cell interactions. The binding of RE-1 polypeptide to rare earth nanomaterials is through specific high-affinity interactions, rather than non-specific adsorption and covalent modification. This binding method has the following advantages: First of all, the peptide synthesis and encapsulation method is simple and easy to operate, with strong controllability. Since the RE-1 short peptide can specifically bind rare earth nanomaterials, the synthesis process is only simple mixing to achieve the purpose of regulating the interaction between nanomaterials and cells. Moreover, polypeptides can be synthesized commercially, so we can synthesize a series of double peptides or polypeptides with multiple functions at any time according to needs. Secondly, RE-1 short peptide has material specificity. RE-1 short peptide can bind rare earth nanomaterials in a broad spectrum, and can specifically bind rare earth nanomaterials with high affinity in many material composites. Meanwhile, the nanomaterials can reduce the sedimentation and sedimentation of rare earth nanomaterials in water, and improve suspension capacity. Therefore, the autophagy of nanomaterials and

Y. Zhang, *Tuning Autophagy-Inducing Activity and Toxicity for Lanthanide Nanocrystals*, Springer Theses, https://doi.org/10.1007/978-981-16-8166-0_7

the toxic effects caused by the nanomaterials can be shielded, and teh biosafety of nanomaterials can also be improved.

Subsequently, the bifunctional compound short peptide RE-RGD-1 composed of RE-1 and RGD sequence synthesized by us can increase the autophagy and toxic effects of rare earth upconversion luminescent nanomaterials in cells that can highly express integrin by interacting with integrins outside the cell. It shows that the targeted strategy is expected to achieve the goals of shielding autophagy in normal cells and improving autophagy in target cells, at the same time, providing a major opportunity for rare earth nanomaterials in cancer diagnosis and treatment, and improving the therapeutic effect of nanomaterials on tumors.

It can be seen that the biological safety problem of nanomaterials can be solved by artificial regulation of the autophagy effect. At the same time, to establish a biosafety evaluation system for nanomaterials to guide and standardize the production and safe application of nanomaterials, systematically study the effect and mechanism of nanomaterials regulating autophagy and its relationship with biosafety, and establish corresponding detection systems and evaluation standards are of great significance.

This article first studied the use of a series of specific surface-binding peptides to artificially regulate the autophagy behavior of rare earth nanomaterials, thereby greatly reducing the toxic and side effects of nanomaterials and improving the killing effect on tumor target cells. The biological safety of nanomaterials can be solved by artificially regulating the autophagy effect, and the effect of radiotherapy and chemotherapy on cancer cells can be achieved by regulating the autophagy level of nanomaterials. This achievement provides new methods and new ideas for the diagnosis and treatment of rare earth nanomaterials in the body.

7.2 The Outlook of RE-1 Application

7.2.1 A Possible Application Case of RE-1 in Detection and Diagnosis

In the field of nanomaterials used in medical diagnostic imaging, contrast agents such as nanomaterials need to reach diseased organs or tissues and stay in this area. For these needs, RE-1 peptide will be the perfect choice. It can not only improve the signal-to-noise ratio of nanomaterials in medical diagnostic imaging, but also reduce toxicity problems caused by non-specific interactions between nanoparticles and cell surfaces. On the other hand, in organ homing or cell localization applications, a bifunctional RE-1 variant peptide linked to a targeting polypeptide is also an ideal choice. We can target and transport fluorescent nanomaterials to the cells, tissues and organs that we want to observe by connecting unconnected cell and organ targeting peptides. After connecting RE-1, it can achieve enhanced specific targeting and reduce non-specific interactions, so as to achieve the purpose of targeted imaging, tumor diagnosis and treatment.

At present, we have synthesized the apoptotic cell targeting peptide APT (CLSYYPSYC) and RE-1 bifunctional apoptosis targeting compound short peptide, which provides a possible application case for the RE-1 short peptide in the detection of apoptotic cells. Earlier experimental results showed that UCN encapsulated with the apoptosis-targeting bifunctional compound short peptide can effectively target the surface of B16 cells treated with DOX, preliminarily marking apoptotic cells by UCN (Fig. 7.1). The success of this research will mark a new field of fluorescence detection and imaging research. This is because the currently widely used fluorescent dyes for detection and imaging are small chemical molecules, such as FITC and Rhodamine. However, these chemical molecules have a common shortcoming, that is, their luminescence time is short and quenching easily occurs. In addition, due to the autofluorescence of tissues and organs, the bottleneck of high background and low signal-to-noise ratio greatly limits the in vivo application of these small chemical molecules when detecting in vivo. As a result, rare earth upconversion luminescent nanomaterials based on their unique upconversion fluorescence characteristics stand out among many fluorescent labeling molecules and show great advantages. The use of this material for in vivo and in vitro detection and diagnosis has the advantages of no background fluorescence interference, strong fluorescence,

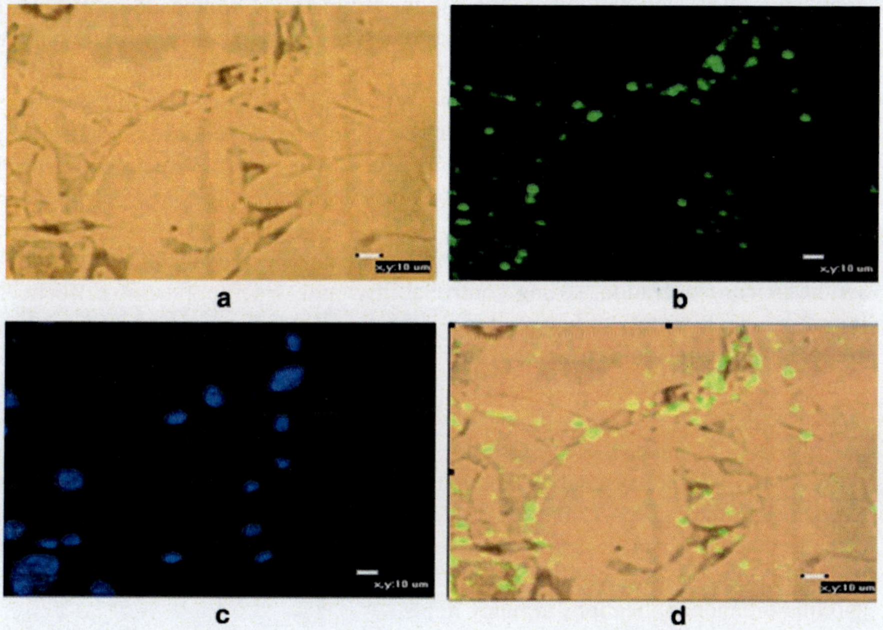

Fig. 7.1 Confocal image showing the detection of apoptotic cells by UCN coated with APT-RE-1 bifunctional short peptide. **a** shows the bright field field of B16 cells after DOX treatment; **b** shows the 980 nm laser channel; **c** shows the nucleus stained with DAPI; **d** shows the overlapping images of **a** and **b**

long luminescence duration, and high signal-to-noise ratio. Therefore, the bifunctional apoptosis-targeting composite short peptide we studied is expected to replace the fluorescence molecular detection method of Annexin V, and provide a new idea and new method for the detection of apoptotic cells.

7.2.2 A Possible Application Case of RE-1 in the Field of New Material Development

As early as 2000, there were reports of using peptides as templates to guide the assembly and synthesis of nanomaterials. However, the use of polypeptides as templates to biomineralize metal or metal compound nanomaterials is limited to the assembly and synthesis of single nanomaterials. As for the assembly and synthesis of new composite nanomaterials doped with multiple materials, bifunctional peptides show unique advantages. We have synthesized a bifunctional composite short peptide CNT-RE-1, which is composed of carbon nanotube specific binding peptide CNT (HSSYWYAFNNKT) and RE-1 short peptide. This peptide can effectively bind carbon nanotubes and rare earth nanomaterials, at the same time, promote the further application of carbon nanotubes. We use the bifunctional composite short peptide CNT-RE-1 to assemble carbon nanotubes and rare earth upconversion luminescent nanoparticles UCN-S, so that the long tubular carbon nanotube surface is assembled with a layer of upconversion luminescence characteristics, which has enabled the successful visualization of carbon nanotubes that are widely used in biosensors, tissue transplantation, and tissue repair. This new type of composite nanomaterials assembled based on bifunctional peptides provides new methods and new ideas for the diagnosis and treatment of diseases.

In future research, the RE-1 short peptide is based on the advantages of simple synthesis, easy operation, strong controllability, and wide application, effectively improving the biosafety of nanomaterials, the signal-to-noise ratio of medical diagnostic imaging, and cancer targeted therapy. It will certainly play an important role in promoting the broader development of nanomaterials and biomedical research. However, there are still some unresolved problems in the application of RE-1 short peptides as small protein molecules in vivo, such as degradability. The mechanism of action of RE-1 short peptide and nanomaterials and the in vivo behavior of nanomaterials encapsulating RE-1 short peptide need further study. Therefore, on the basis of comprehensive and in-depth research on the mechanism of short peptides and nanomaterials, we must eliminate difficulties to solve problems, develop vigorously, and find broader applications in many fields.